ervices

elow.

SUSTAINABLE RURAL DEVELOPMENT

Sustainable Rural Development

EDITED BY

HENK DE HAAN
Department of Sociology
Wageningen Agricultural University
Wageningen, the Netherlands

BABIS KASIMIS
Department of Economics
University of Patras
Agrinion, Greece
Institute of Urban and Rural Sociology
National Centre for Social Research
Athens, Greece

MICHAEL REDCLIFT
Wye College
University of London
Wye, UK

Ashgate

Aldershot • Brookfield USA • Singapore • Sydney

Published by
Ashgate Publishing Limited
Gower House
Croft Road
Aldershot
Hants GU11 3HR
England

Ashgate Publishing Company
Old Post Road
Brookfield
Vermont 05036
USA

British Library Cataloguing in Publication Data
Sustainable rural development
1.Sustainable development 2.Rural development
I.Haan, Henk de II.Kasimis, Babis III.Redclift, Michael, 1946-
307.1'412

Library of Congress Catalog Card Number: 97-72077

ISBN 1 85972 595 3

Printed and bound by Athenaeum Press, Ltd.,
Gateshead, Tyne & Wear.

Contents

Figures and tables

List of contributors

Evi Arahoviti, Department of Agricultural Economics, Agricultural University of Athens, Athens, Greece.

Jean-Paul Billaud, CNRS, Paris X University, Nanterre, France.

Karl Bruckmeier, Katalyse Institute, Cologne, Germany.

Hugh Campbell, Department of Anthropology, University of Otago, Dunedin, New Zealand.

Mary Cawley, Department of Geography, University College Galway, Galway, Ireland.

Desmond A. Gillmor, Department of Geography, Trinity College, Dublin, Ireland.

Henk de Haan, Department of Sociology, Wageningen Agricultural University, Wageningen, the Netherlands.

Ina Horlings, Centre for Agriculture and the Environment, Utrecht, the Netherlands.

Babis Kasimis, Department of Economics, University of Patras, Agrinion, Greece; Institute of Urban and Rural Sociology, National Centre for Social Research, Athens, Greece.

Leonidas Louloudis, Department of Agricultural Economics, Agricultural University of Athens, Athens, Greece.

Perpetua McDonagh, Department of Geography, Trinity College, Dublin, Ireland.

Berit Nygård, Centre for Rural Research, Norwegian University of Science and Technology, Trondheim, Norway.

Giorgio Osti, Department of Social Sciences, University of Trieste, Trieste, Italy.

Teresa Patricio, CIES-ISCTE, Lisbon, Portugal.

Florence Pinton, Paris X University, Nanterre, France.

Michael Redclift, Wye College, University of London, Wye, UK.

Oddveig Storstad, Centre for Rural Research, Norwegian University of Science and Technology, Trondheim, Norway.

Hilary Tovey, Sociology Department, Trinity College, Dublin, Ireland.

Alenka Verbole, University of Ljubljana, Biotechnical Faculty, Agronomy Department, Ljubljana, Slovenia; Wageningen Agricultural University, Centre for Recreation and Tourism Studies, Wageningen, the Netherlands.

Introduction

The European Society for Rural Sociology held its 16th congress in Prague in August 1995. The congress theme, 'From productivism to sustainable rural development? The transformation of rural economy, society and space in the late 20th century,' expressed the importance of environment-induced change in rural areas and attracted many participants from various backgrounds. The congress revealed a remarkable convergence of research in rural studies. If theoretical and political controversy was characteristic for the period of 'classical' modernization, there is now broad political and scientific consensus about such concepts as 'sustainability,' 'ecological modernization,' and 'quality.' Although debate about the proper forms of regulation continue and theoretical controversies about what is to be sustained persist, it is now widely recognized that productionist models have resulted in agricultural surpluses, aggravated budgetary problems and imply serious social, cultural and environmental consequences for rural areas. The crisis and demise of the old regulatory framework has led to rural restructuring with new forms of producing and consuming rural goods. The previous focus on agriculture has moved towards a much broader perspective on the relation between society and rural resources, including wider concepts of production and consumption. 'Sustainable development' has become a central element in current discourses on rural development, modelling perceptions of past and future trends.

This book publishes a selection of papers that were presented at the Prague congress in 1995. They illustrate the spectrum of current research interests on sustainable rural development. The papers range from a description of national discourse about the rural environment to detailed local studies of changing farmers' culture; and from analysis of food systems to strategies at the farm level. A number of chapters explicitly discuss the concept of sustainable development and its application to rural society, while others confine themselves to specific geographical locations, fleshing the topic out through empirical research in Ireland, Slovenia, Greece, the Netherlands and New Zealand.

The fact that the concept of 'sustainability' precipitates such a wide variety of empirical studies and theoretical reflection illustrates that it is both concrete and ambiguous. As Billaud, Bruckmeier, Patricio and Pinton show, the concept has its antecedents in various national traditions, and the way in which different cultures perceive and discuss it reflects traditional preoccupations with the peasantry, rural social life, landscape and nature. The concept of sustainability acquires a variety of meanings and represents a range of political goals in different contexts, and at different levels. Tovey's paper nicely illustrates the multiple character of both traditionalism and modernism; it seems as if the present concern with sustainability will also result in differential paths of development, reflecting both continuity and transformation.

A century ago few people would have thought rural societies and economies were anything but sustainable. The modern agri-food system had not developed in Europe and North America, most people were still living in rural areas, many of them dedicated to the kind of activities undertaken by countless generations before them. Rural societies, in Europe as in much of the rest of the world, were essentially agricultural societies, and most non-agricultural employment was closely linked to farming, in one way or another. But if the rural societies of a century ago were sustainable, why were people leaving them in their millions? Why were they moving to cities in Europe, or settling in the Americas and the European colonial states of Asia and Africa? This is a simple reminder of the dangers of projecting our aspirations for the present and future onto the past.

The discourse on sustainability is not only inspired by history, tradition and current concerns about the environment. The idea of sustainability has particular attractions for students of rural society because of its close association with nature and biological systems. In plant ecology, for example, the concept of 'sustainability' is associated with successional change in plant communities, and later became a *model* for the management of forests and rangeland. The concept of sustainable yield, which pre-dates the much wider discourse about sustainable development prompted by the Brundtland Commission in the 1980s, has its origins in forestry and fisheries' research. It echoes the idea of ecologists such as Odum that, if we expect ecological systems to produce flows (of goods or incomes), then 'evolutionary adjustment' on the part of the ecosystems subjected to rapid change would also be required.

From an ecological point of view, *agricultural systems* should demonstrate the capacity to renew themselves (regenerate) in the face of disturbances (stress and shocks). 'Sustainable utilization,' the watchword of the first *World Conservation Strategy* in 1980, took up the ecological notion of resilience in the face of externally generated change and applied it to human-induced changes, to planning and 'development.' Soon sustainable development was to emerge as a fully normative concept from the chrysalis of natural science.

The idea of sustainable development has been transformed by the social sciences from a model *of how natural systems work* to a *model of how social systems might work*. Nowhere has this more important implications than for rural sociology. Social science arose during the nineteenth century from the fusion of rapid industrialization, the revolution in ideas marked by the French and American Revolutions, and the unprecedented development of cities. Social science in general, and sociology in particular, sought to draw attention to the social and economic problems associated with these traumatic changes. Before long, however, social scientists began to work with and inside government, sharing responsibility for the new social order that they continued to criticize. Most of the major planning exercises targetting rural society, in developing as in developed countries, take the idea of managing change in the interest of communities and social groups within rural societies themselves as their point of departure. This is the legacy of rural sociology, which is very closely allied with public policy, in an increasingly transnational context.

As the contributors to this volume demonstrate, the environment and sustainability represent at least a covert challenge to this orthodoxy of public policy as a reflection of the interests of rural society. There are clear discrepancies and contradictions between interests within rural society, not all of which support 'sustainable' objectives in anything more than a rhetorical sense. Transnational corporations in the food industry often meet opposition from the advocates of organic agriculture, but may also seek to advance their commercial interests by espousing many of the same objectives. The restructuring of agri-food networks cannot be separated from the transnational policies, which are often represented as measures to contain the interests of the food system itself.

If we want to identify the importance of sustainability for rural societies, we clearly need to look beyond the stated goals of governments and the European Union, to the *institutions* of rural societies. We do not merely bequeath our environment to future generations; we also bequeath the institutions with which future generations will manage it. Most of the papers in this book demonstrate the ambiguity which surrounds these institutions, which frequently represent both an obstacle to sustainability, and a challenge in seeking to achieve it. For rural sociologists the primary concern today is with what is socially sustainable in the new institutional landscape afforded by the late twentieth century.

Part I contains two chapters with general overviews by Billaud, Bruckmeier, Patricio and Pinton, and Redclift. Chapter 1 explores the way in which the rural environment is socially constructed, its premise being that human societies define the idea of sustainability. Billaud et al. argue that European agri-environmental policy faces problems of social acceptance and legitimation. 'Europeanization' has not yet created a new supra-national cultural reservoir. The concepts and discourses on rurality, agriculture and the environment are reframed by national cultural traditions and interests. Chapter 2, by Redclift, looks at the way in which industrial and 'natural' systems converge around agriculture, and serve to transform the environment. It opens up questions about what we mean by sustainable levels of resource exploitation in rural societies. The first two chapters thus pose a series of questions about human behaviour and sustainability. What prevents precautionary action from being taken in rural societies? Which are the operative networks of social communication? How are environmental issues socially channelled in rural areas? How are consent and compliance negotiated?

Part II on globalization and the food system examines the national and international contexts in which the agri-food system is developing, and the implications it carries for farm production and food consumption. Primary agricultural production still occupies most of the space in rural areas. The introduction of ecologically-friendly farming methods, resulting in sustainable forms of rural resource use and healthier food, is not, however, determined by farmers alone. Most farmers are integrated into complex chains of input industries, food-processing, wholesalers and retailers, often organized on an international scale. Changing consumer demands can only be met by breaking out of this chain, as is often the case with organic farming, or if demands for quality food are channelled to them via the food and retail industry. Nygård and Storstad describe changes in food distribution in Norway, showing the emerging struggle for power over distribution and decisions about what consumers will find in the shops. The outcome of this struggle will obviously exert considerable influence on the chances of successful ecologization of production chains. The relevance of markets and the influence of large companies in the food system is clearly demonstrated by Campbell. Organic agriculture in New Zealand is appropriated by big firms, linking farmers in order to meet the growing demands for organic food on the Japanese market. The restructuring of the vegetable commodity system towards more sustainable forms has created an opportunity for many conventional farmers to enter organic production for the first time. These chapters raise the question of the appropriate level at which the development of sustainable agriculture should be analysed. It shows that market-led and food system induced changes in forms of production may well be as important as state regulation and farmers' own initiatives.

Part III presents case studies from Ireland, Greece and the Netherlands, on agrarian change and the environment. In Chapter 5, Cawley, Gillmore and McDonagh analyse the adoption of alternative farming enterprises by Irish farmers in response to over-production and reduced incomes. Enlarging on-farm income possibilities certainly is not only important for the survival of small family farmers, but also of vital importance for the rural economy and society in less-developed areas. Maintaining the population on the land is a pre-condition for the very existence of rural society and thus for sustainable rural development. Louloudis and Arahoviti (Chapter 6) approach the persistence of rural society from another angle. Many parts of the Greek countryside are under heavy pressure from tourist development and urbanization, resulting in a drastic reduction of agricultural land. The authors argue that the new European agro-environmental measures will remain largely ineffective as long as the legal protection of agricultural land remains unresolved. Their paper is a good illustration of the discrepancies between national and European policies. The full implications of this process are only beginning to be apparent in much of Southern Europe. Tovey, in Chapter 7, raises the question whether sustainable agriculture is only possible if it is based on farmers' traditional culture and local identity. The revival of craftsmanship and an ethos of stewardship are often associated with the revitalization of locally embedded farming systems, both in academic and political discourse. The assumption is that the productionist ethos has largely eliminated the cultural and social basis for an alternative outlook on production. Tovey argues, however, that Meath farmers in Ireland have no common historical identity and no connection with the traditional peasantry. Instead, they are oriented towards the future and much engaged in professional, entrepreneurial practices. They nevertheless consider environmental concerns as part of their occupation, not because they have themselves been converted from 'productionist' to more ecological orientations, but to prevent control over the countryside from slipping into the hands of non-farming or urban groups. Horlings takes a somewhat different perspective on changes to the farming profession in Chapter 8, although she too concludes that farmers defend themselves against forms of regulation that diminish their control over the labour process and the management of natural resources more generally. Dutch farmers have been subject to a long-term process of de-skilling and growing dependence on external material and scientific resources, according to Horlings. She argues for substantial forms of freedom for farmers as a necessary condition for the reintroduction of farming skills and craftsmanship. She supports her argument with a case study on 'innovation groups': farmers who create alternative perspectives and design their own solutions to the current environmental problems. Their initiatives are analysed as endogenous contributions to a more sustainable agriculture. Tovey and Horlings raise the question of human response to changes in markets and regulation. Sustainable development needs to incorporate rural peoples' logic and conscious behaviour, at least as a planning goal, hereby acquiring a measure of 'reflexivity.' The importance of this logic *vis-à-vis* other actors in the agri-food system illustrates the multi-levelled character of sustainable development.

Part IV on regional development takes the issue of rural sustainability into the European periphery. Osti's chapter, which deals with mountain regions in Italy, shows how difficult it is to reconcile the human and 'natural' dimensions of sustainability. He shows that mountain areas are unable to rely on indigenous development, however much some advocates of sustainable development insist upon its desirability. Mountain areas today are beset by what he terms 'a revival of the myth of progress,' representing uncontaminated enclaves of civilization, for tourists and planners. Such attitudes demonstrate a gulf between what Osti sees as the 'models' of development, and the real 'processes' which characterize the rural periphery. Finally, Verbole's chapter looks at the advantages and disadvantages of rural tourism. This contribution, based on material from Slovenia, poses the problem of endogenous and exogenous development in a very different way from the previous chapter on Italy. Verbole argues that ". . . Sustainable rural tourism development in Slovenia should . . . be based on defining and activating local development to realize the indigenous potentials of rural areas in all their geographic, cultural, and natural diversity." Behind the surface of regional development in rural areas is a potential, in Verbole's view, for a more autochthonous type of development, development from 'inside,' that does not simply locate the region within the broader European canvas.

This book is the tangible product of international cooperation, from three editors, Dutch, Greek and British, and its contents demonstrate the way in which all our futures are intertwined, particularly in modern Europe. The contents also demonstrate the vigour and versatility of much contemporary rural sociology, as was clearly evinced by the European Society for Rural Sociology congress in Prague in the summer of 1995.

Part I

General Overviews

1 Social construction of the rural environment. Europe and discourses in France, Germany and Portugal

Jean-Paul Billaud, Karl Bruckmeier, Teresa Patricio and Florence Pinton[1]

Introduction

The European Union is partly a political order and partly a social system. The common market and the supra-national political order do not represent the autonomy or self-organization capacities of states and societies.[2] This emerging European order is also not a new supra-national cultural order in its present stage of evolution, but through its interaction with national political and social systems it influences social discourses which have been shaped in national or regional cultural traditions and are still framed in these traditions. For example, the Common Agricultural Policy and the rapidly expanding environmental policy of the EU are influencing social perceptions and discussions of countryside and rural problems. This influence is political, but it may not be limited to political institutions and political decision making. Furthermore, its impact is not well known, although there is growing research and literature on European environmental policy and on the consequences of European political integration.[3]

Instead of asking how a genuine European order and discourse is evolving above the national societies of its member countries, our question is the other way round: How does the emerging European order, whatever its quality, change the ongoing discourse on rural society, agriculture and environment at national level? By 'national discourse' we mean not only coherent and consensual national ideologies of rurality which are vague and symbolic, but also forms of discussion about countryside, agriculture and environment which can be traced in the history and culture of the respective country. The notion of discourse indicates that there is always a process in which varying ideas about social order are discussed more or less controversially. Such discussions are conducted at various levels in everyday life, the media, political debates and science, and they draw in a variety of social groups, organizations and actors. Discourses at national level use certain socio-culturally codified traditions of rural society and agriculture. The structuring principle of the discourse can be seen in policies and political controversies concerning countryside, agriculture, and environment. The cleavage lines along which these discourses split are different from one country to another. However, they can be interpreted as variants of a confrontation between tradition and modernity in abstract terms – in most cases a confrontation between agricultural and industrial types of development.

We compare national traditions of discourse on the rural environment in three culturally, socially, economically and geographically distinct countries: France, Germany and Portugal, and note how these discourses are changing under European influence.[4] The selection of these countries is not purely systematic, but due to the contingencies of the specific research programme and the fact that other projects have done similar research in other European countries.[5] The dynamics of European integration as an interaction between European and national systems runs like a thread through the comparison.[6]

The reason why environmental problems of non-industrial origin should have arisen in the countryside, seems straightforward enough when the impact of modernization and intensification of agriculture, at least in the northern member states of the EU, is considered. The question of how the relationship between countryside and environment is perceived socially, cannot be answered by simply referring to the modernization process nor by a logical explanation logic of the kind 'first technical and economic modernization, then environmental pollution, then social and political perception of the problems, then environmental action.' Between EU member states there are differing socio-cultural time frames, 'logics of development' and codifications of these issues. Several national examples need to be compared to point up such differences. We start with a review of the main elements of traditional social construction of rurality in the three countries. Then we sketch the agri-environmental discourses in the framework of environmental policies since their inception in the early 1970s. Finally, we discuss the role of the agri-environmental programmes of the EU and their significance for national discourses on rurality, agriculture and environment.[7]

Traditional social constructions of rurality and the modernization process

Traditional social constructions of rurality refer to sets of ideas or 'paradigms' of rural life which originated before the modernization of European agriculture in the twentieth century and reflect a pre-industrial state of rural life. Such paradigms date back mainly to the nineteenth century when industrialization had already begun. These traditional constructions are forms of social thinking which have survived until now in competition with ideas about industrial modernization. The environmental discourse gives a new momentum to such traditional constructions, which, combined with agriculture, cluster in ideal-type configurations like 'traditional extensive agriculture socially organized as village-based family farming.' These constructions are seldom the creation of one mind. They tend to be produced by a multiplicity of social groups and actors in rural and urban area and reflect only a vague symbolic consensus on the role of rural areas and rural, mainly agricultural, economy in the broader society. The notion of traditional social constructions of rurality allows us to describe rural traditions and historical developments at national level, which makes it possible to reconstruct present agri-environmental discourses. In describing such traditions we find that social constructions of rurality are not changing as rapidly as the social structures existing in modern European societies.

In France, Germany and Portugal we find notions of rurality which are based on the dichotomy of urban and village life, with the peasants being the traditional social group for the development of rural society, and agriculture the dominant economic activity. However, the history, development and cultural codifications of this dichotomy have taken rather different paths.[8]

France

In France, with its strong agricultural tradition, the notion of countryside or rural life is based on the radical opposition of town and village, which has been in operation since the end of the nineteenth century. The peasants were the dominant social group in the evolution of the countryside, not only for demographic but also for political and ideological reasons.[9] In spite of the modernization of the countryside with the subsequent reduction of the peasantry to a minority social group, the 'rural' remains assimilated with agriculture and synonymous with stability and tradition. Simultaneously a process of negation and re-creation of the peasantry occurs, with limited reference to the social roles of the peasantry or by making the peasant into a mythical or idealized figure.

The rural plays a fundamental role in the representation and categories used to rethink French social development. It re-emerges during times of social crises. Fear of the workers' movement and of urban revolution places the rural in the centre of the ideological and political game during the Third Republic. The deep identity crisis provoked by the defeat in the 1940s gives rise to an agrarian movement identified with the 'return to the land.' And finally, the crisis provoked by an intensive process of modernization in the post-war period awakens a longing for a 'rural utopia.' From this historical development, the main elements of a notion of rurality focusing on traditional rural life and its change through modernization, can be abstracted as sets of contradictory notions about rural life:

- The social dominance of agriculture and the peasantry is fixed in one set of ideas, although peasants and agriculture have been numerically exceeded by other professions (workers) and other economic activities in the countryside;
- The radical transformation of agriculture, its integration into the national economy, and the orientation to modernity is not reflected in perceptions of the countryside; even today rural life is often synonymous with stability;[10]
- Simultaneously with this ideological social codification of the peasantry unfolds a notion of rurality without reference to the social roles of peasants. This is the converging point of positive alternatives to urbanism (village community, direct production in relation with nature, local markets), or in the notion of the countryside as 'green lungs' for urban society.

Among the various alternative movements and ideological counter-attacks on industrialization in the sense of 'back to the earth,' celebrating rural life, the neo-rural movement of the 1970s is the latest, creating a new rural utopia, born out of the crisis of urbanism and industrialism. It is this movement which combines the issues of rurality, agriculture and environment (Mathieu and Jollivet, 1989; Viard, 1990) and finally leads to a broader discussion of the identity crisis of the agricultural profession.[11]

Germany

In Germany, with its later process of industrialization, the separation of rural and urban life has given rise to various ideologies derived from this dichotomy which became influential since the late eighteenth century.[12] The intellectually based constructions of rural life and the countryside transmitted to us in cultural artifacts are based on the perception of non-rural, urban dwellers and intellectuals on the eve of industrialization. Literary constructions of rurality, rather than anonymous oral traditions, form the roots of a social concept of rurality, which during the nineteenth and twentieth centuries has often been ideologically overplayed. The social organization which unfolded within traditional agriculture in the nineteenth century as part of a traditional social construction of rural life is that of the peasant family farm. This gave rise to an influential social ideology which still remains of importance for agricultural policy.[13]

Agrarian romanticism, rooted in the ideological confrontation of a 'natural and sound rural life' with the opposite connotations of urban life, was reflected sociologically at the end of nineteenth century: the urban–rural dichotomy reappeared in the Toennies' concepts of 'Gemeinschaft' and 'Gesellschaft.'[14] Politically the doom of agrarian romanticism can be seen in its merging with the Fascist ideology of peasant-based agriculture.

In the traditional concept of rural life the notion of environment is represented by nature. Before the analytical notion of environment became important in science and politics in the twentieth century, nature was synonymous with this term, signifying 'our non-human surroundings' (Simmons, 1993, p. 11). Additionally, the notion of landscape has become a semantic representation of the term environment.[15] In the socially produced form of landscape which is called 'agrarian landscape' (*Agrarlandschaft*) or 'cultural landscape' (*Kulturlandschaft*), we can observe the evolution of social interaction with nature as based on agriculture, whereas industrialization produces artificial landscapes characterized by technical infrastructures and artifacts. The traditional structures of agrarian landscapes have managed to survive partly until this day (Born, 1989; Küster, 1995). It is the rapid modernization process of agriculture combined with farmland consolidation (in East and West Germany under differing social systems) which has been responsible for the gradual disappearance of these differentiated landscape structures since the 1960s, and this has accelerated criticism of modern intensified agriculture, emerging in the 1950s, for creating uniform landscapes. In brief, traditional social constructions of rurality are retained in the notions of 'cultural landscape' and 'peasant family farming.' It was with the emergence of environmental movements criticizing modern agriculture that the values combined with these ideas began to give way.[16]

Portugal

In Portugal, the rural–urban dichotomy still remains a central feature of the social construction of rurality. The traditional construction can be traced to the important role played by agriculture in the economy and in maintaining land use. Equally important in shaping the social construction of rurality has been the isolation of Portuguese rural society throughout the nineteenth and most of the twentieth century. The late process of industrialization was largely restricted to the urban coastal areas and most of rural society retained its structure of small traditional family farms, with the exception of the latifundia in the Alentejo. While rural society provided the necessary labour force for the urban coastal areas during the process of industrialization, it remained isolated and undeveloped. Modernization of agriculture, where it took place, was restricted to very specific areas. Throughout the process of industrialization rural society provided a crucial element of social and ideological continuity. Rural society became idealized as a focus of tradition and necessary for the preservation of tradition. It became a safe haven, a refuge against the evils and ills of modernization. Isolated and sheltered from the rest of the world rural society became associated with such traditional family values as simplicity, obedience, religiosity, discipline and hard work. The values associated with rural society provided the ideological basis for fifty years of non-democratic rule.

The traditional farm survived through various combinations of family strategies. Alternative forms of family employment and income allowed subsistence farmers to survive, while providing the countryside with different forms of production. Farming for many families became one of several different economic activities. Mass emigration to western Europe during the 1960s did not lead to the abandonment of the countryside. On the contrary, emigration provided an extra source of income as emigrant remittances assured the remaining family members on the farm that the property would continue in their possession and that land would be in use.

The social construction of rural society links agriculture with other forms of economic activity and largely includes an active rural community. In the absence of modernization of agriculture, the rural population was obliged to seek alternative sources of income while keeping land in cultivation. Farmers seek alternatives to agriculture so that farming can be maintained. Agriculture is an important part of rural society but it is seen as insufficient for maintaining rural society.

After the long period of colonialism, isolation from Europe and political dictatorship in the twentieth century, traditional rural society and the dominance of agriculture in Portugal came to an end only in the 1970s and 1980s with the democratic revolution and membership in the EC.[17] The importance and weight of the rural population for Portuguese society in general has been largely responsible for the dominance of a strong rural emphasis within social and public discourse (Almeida, 1994), including scientific discourse.[18]

Three nations compared

In all three nations with their traditions of rural life we find varying forms of dualistic concepts of society. Dichotomies of village and city, traditional and modern economic sectors, local and national or global cultures can be seen (Pongratz 1988). These are elements of social constructions which are co-existing in different forms. In France, the social taboo against criticizing agriculture is mainly based on traditional constructions of agriculture, which are revitalized in critical phases of development where disruption between tradition and modernity results in an impasse. In Germany traditional and modern elements of rurality co-exist as social constructions and differing forms of socio-economic development. In Portugal the dualism unfolds as one of internal (national) and external (EC-influenced) development. Within the countries the social discourses on rurality and environment vary according to the social and political actors involved (peasants, governments, political groups, various intellectual movements, different scientific disciplines) and levels (local and national), at which the constructions are used with the local community being the original form of organization and self-perception of peasant cultures. These variations cannot be described in greater detail here. What has become manifest from this short review of national traditions is that these constructions are anchored in heterogeneous cultural identities which need to be explained by history.

The literary discourses can be divided into cultural-normative and scientific. Each nation had 'leading ideologies' of rural life and agriculture valid for certain periods. In France such ideologies, since the late nineteenth century, seemed to result from the economic significance and social esteem of agriculture and the peasantry. In Germany the notions of 'agrarian romanticism' used by certain intellectual groups, conservative groups and also the Fascist movement had some influence. In Portugal important facets of such rural ideologies seem to be religion and family-bound values of rural life, and more practice-bound social constructions resulting from the tradition of 'action research,' with ethnographers and other scientific and technical experts intervening in rural life.

Although the notion of environment is absent from all traditional concepts of rurality, the environmental connotations of traditional social constructions of rurality can be identified. These connotations are based on the dominant role of agriculture in rural areas. They can be split into parts of the landscape which are shaped under the influence of agriculture, and other parts which represent the non-appropriated and non-used parts of nature (nature protection, national parks etc.). The relationship between agriculture and environment is reflected in notions like high quality of agricultural products and food, agrarian landscapes and natural habitats, rural communities as a form of integration of nature and society, maintaining the landscape and habitats. Beyond agricultural production, the environmental dimensions of rurality are expressed in normative concepts related to recreation, e.g., the beautiful landscape or 'free nature' which have become a romantic or utopian orientation for urban dwellers and intellectuals.

Development of agri-environmental discourses in the framework of environmental policies

The political shocks of 1968 in western Europe mark the end of post-war social development and the beginning of new social conflicts and a new environmental discourse as a critical reaction to modernization. This includes agricultural modernization in its productivist forms under the Common Agricultural Policy of the EC since the 1960s. In France and Germany this new political and social discourse on the environment began in the years after 1968, with governmental programmes. Portugal, at this time still under dictatorship, was outside the mainstream political development in western Europe and, like Spain, isolated from European development.

France

In France, the programmes of the government of Chaban-Delmas since 1970 and the foundation of a Ministry of the Environment (although not very powerful and not well integrated into French state administration) in 1971 indicate the beginning of a new political development. However, the state budget for environmental policy remained low compared with other industrialized countries. In 1974 a first ecological candidate (René Dumont) appeared on the scene, and in 1977 green candidates stood for the municipal elections. The agricultural sector was excluded from this new environmental discourse, which focused on industrial problems.[19] Also the political programmes and measures directly addressing nature protection in the countryside do not touch agriculture or agricultural land use; they refer to non-agricultural areas only (forests, national and regional parks, natural reserves). Only the oil crisis in the mid-1970s led to a short-term political appraisal of agriculture as producer of 'green oil.' Until the end of the 1980s a critical discussion of the relationships between modern agriculture and environment was a political taboo in France, environmental problems being seen as exclusively urban and industrial by the majority of society. When the EU introduced agri-environmental measures in 1985, the reluctance of the French government to implement such policies can be explained by:

- the importance of the agricultural sector in the French national economy (the biggest agricultural producer in the EU, the world's second biggest exporter of agricultural products);
- the importance of the farming lobby in relation to the management of rural landscapes (well organized, with strong influence on the government);
- the importance of the peasantry as an ideological component in national identity.

Even with the activation of a critical agri-environmental discourse since the late 1980s, the problems of agricultural pollution have not been addressed directly but mainly via abstract scientific concepts and debates on

- risk (from risks associated with reduced grassland and risks of forest fire to risks caused by intensification;
- biodiversity (stimulated by the demarcation of environmentally sensitive areas);
- landscape management (from the viewpoint of scenic landscapes for tourism as opposed to the production-oriented land-use planning serving agriculture). From this issue a specifically French construction is derived – that the peasant must remain the privileged manager of the landscape. This is a contested issue between peasants and other social groups, and environmental concern undermines the traditional legitimation of the peasant as a producer of food for the whole society.

Germany

In Germany, the social-liberal coalition government of Brandt (in power since 1969) opened the new phase of environmental policy. The first state environmental programme was introduced in 1971. Environmental policy unfolded in a series of phases (Bechmann, 1988) parallel to the unfolding of the environmental discourse in German society. This close interaction between government policy and social discourse leaves the impression of a rapid institutionalization of ecological protest. The opening phase was formulated by the federal government, whereas in the subsequent phase from 1975 to 1982, critical environmental movements and the Green Party were involved. The political climate became highly conflictual in these years, as can be seen in the anti-nuclear protests.[20] Between 1983 and 1990 environmental protest and movements became institutionalized (e.g., first federal state governments in which the Green Party took part) and ecological orientations became widespread throughout society. Only during this third phase of the environmental discourse did the sub-discourse on the rural environment gain momentum, with a critical discussion of agriculture and its environmental impacts.[21] This is expressed by the development of an agrarian opposition movement in the 1980s (which united parts of environmental and Third-World movements with small groups of farmers critically aware of the environmental threats of modern agriculture) and, based on a multidisciplinary scientific analysis, in the first extensive study of the environmental impacts of agriculture (SRU, 1985). Meanwhile this critical discussion has been generalized. The breach within the social organization of agriculture which became manifest within the critical environmental discussion in the 1980s can be seen as a 'silent revolution.' Within the framework of continuing small-scale agriculture based on 'peasant family farming' (the term in use in agricultural policy), a modernized and productivist agriculture had developed. The disruption between the traditional social construction of agriculture and its modern economic and technical base was not reflected socio-culturally until the environmental discourse brought this to the public attention.[22] The end of the separation of East and West Germany and the breakdown of socialism in 1990 defined the environmental situation completely anew. From now onwards East German agriculture, rapidly transforming into a large-scale industrialized sector operating under market conditions, represents the new factor in rural development, devaluing the West German development paradigm based on small- and middle-scale family farming.

Portugal

In Portugal the agri-environmental discourse was not framed in environmental policies for a long time. The absence of modernization of agricultural production helped consolidate a general notion that agriculture does not harm the environment. The dominant perception is of harmony between agriculture and the environment. The general view is that rural society has been preserved because industrialization failed in rural areas. Farmers appear as the natural protectors of rural society – taking care of the land, maintaining it in use, building and rebuilding terraces, clearing the ground, and performing other conservation tasks. Traditional agriculture, as characterized by small subsistence family farms, has preserved rurality and, equally important when considering environmental concerns, maintained land in use. As traditional family farming survived through multiple strategies of preserving property and maintaining land use, it has now provided a fruitful basis for the renewal of rural society in response to demands raised by new environmental concerns.

While Portugal is a relative newcomer to environmental concerns in rural society, some measures dealing with landscape protection and the protection of animal species, and forest growth and protection, have been introduced in the last ten years. Portuguese 'awakening' to environmental issues dates from the mid-1980s with a few legislative measures regulating waste disposal and creating animal- and wildlife-protection areas. Protection of environmentally sensitive natural settings is particularly important to specific areas such as certain biotopes (for example, Castro Verde in the Alentejo), and in wildlife sanctuaries.

Two major orientations are expressed in Portuguese legislation during the 1980s: addressing urban and individual sources of pollution and protecting nature-sensitive areas and wildlife. Finally, in 1990 the Ministry of the Environment and Natural Resources was created. In spite of these recent trends, modernization combined with rapid rates of growth has had a negative effect on the environment and, significantly, generally failed to involve or mobilize the population to its defense.

Only in the 1990s has agriculture begun to be acceptance as a polluter. In spite of the gradual recognition that agriculture carries some blame for damage and problems in the environment, the dominant discourse remains one of relative harmony between agriculture and the environment. This continues to be expressed by the dependent relationship of one upon the other – for the natural environment to be preserved, agriculture must be protected.

A new direction of environmental concerns emerged with Portugal's integration in the European Union. Late European integration created pressure to identify and address agro-environmental issues and, through the implementation of directives and regulations Portuguese agriculture faced increasing pressure. Paradoxically, this pressure pointed in two contradictory directions – on the one hand, to modernize its agricultural systems of production, on the other hand, to reconvert to extensification. A split in the agri-environmental discourse became evident: an 'imported' EU discourse versus the more prevalent, harmonious relationship between agriculture and the environment. The agri-environmental discourse in Portugal is still largely linked to the cultural heritage found in the countryside. The general aim of the agri-environmental measures introduced in Portugal is to preserve traditional systems of agricultural production by combatting social and demographic trends towards desertification. The bulk of agri-environmental measures can be seen as a subsidy toward the preservation of rural society by maintaining land in use.

Discourse on the rural environment compared

European environmental policy, shaping the first critical notion of the relation between agriculture and environment, has also had a catalytic effect on the formulation of a notion of positive relationships with reference to traditional agricultural structures and values. The discourse reveals the systematic weaknesses in an effective environmental movement or national 'green' consciousness. Ecological and environmental problems are blamed on urban and industrial pollution and industrial waste and on pollution at international level.

With the exception of Portugal, France and Germany seem to represent differing examples of the mainstream of agricultural development under EC influence since the 1970s. The beginning of the new environmental discourse is more or less simultaneous, at national, European and global levels, as the 1972 'wake up conference' of UNCED in Stockholm symbolizes. Critical discussion of the rural environment began later, unfolding in the 1980s, in Germany earlier than in France. The specific role of agriculture determines the nature of the discourse on the rural environment too. Until now the countryside is seen mainly as a 'product' of agriculture mainly, although agriculture is marginal in terms of gross national product. Landscape is the framework for integration of land use and environmental quality, with traditional agricultural landscape structures now becoming elements of protection and conservation policies. In Portugal the discourse unfolds later but goes on more rapidly. Certain elements which have been found in the other countries emerge here too – for example, the role attributed to farmers as the main managers of the rural landscape.

A positive relationship between agriculture and environment, which is revitalized in the agri-environmental debates at political levels, especially by such movements as organic farming, dates back to pre-modern agriculture. It refers to traditional agriculture based on a self-sufficient local economy. Traditional agriculture becomes a counter-paradigm to industrial agriculture and a model for ecological modernization. Its basic elements re-emerge in extensive agriculture, organic farming, or (continuation of) traditional relics of agriculture – elements which represent the agri-environmental policy of the EU.

The role of agri-environmental measures of the EU

EC environmental policy began in the 1970s with a series of directives and regulations to attain a common level of environmental quality in the member countries. Although agriculture had been touched by certain EC environmental programmes and measures, for example the directives on drinking water quality, there was no politically conscious and socially perceived EC agri-environmental policy. Until the late 1980s, the CAP was characterized by environmental non-intervention. Only with the manifest crisis of over-production and the introduction of quota regimes in the mid-1980s, was a turning point reached. Agri-environmental policy rapidly became a significant instrument for the reduction of agricultural surpluses.

France

In France the introduction of agri-environmental measures in 1991, to be known as 'article 19,' marked a disruption in agricultural policy. Since the 1960s the effects of modernization and intensification of agricultural land use had been taken into account, but only to compensate for structural disadvantages, not for reasons of environmental protection. Also the local regulation of water quality which began at this time did not touch agriculture and agricultural policy. Article 19 became the symbol for a new agricultural policy (Alphandéry, 1994; Billaud, 1994; Deverre, 1994) and regulation 2078/92 a continuation of this policy. Four experimental sites which were demarcated in 1989 still were of exceptional character. Meanwhile many such zones exist, their multiplication being mainly a consequence of the 'plans of sustainable development' (PDD) in operation since 1991.[23] These have been designed by an inter-ministerial committee to integrate production systems and environmental production and to enlarge the social basis for agri-environmental policy in France. Within three years the original budget increased six times and the number of local projects flourished. The obvious success means that the original objective of the measure to protect the environment is slowly being transformed into a measure of support for marginalized agricultural regions. The National Farmers' Association (FNSEA) also accepted the measures, thus contributing to the emergence of a dual image of agriculture which abandons the traditional French agricultural model with the argument 'while efficient and productive (though less polluting) businesses continue to be socially valued, so is a new form of agriculture that maintains the landscape and produces "new" subsidized services and quality regional food products.' (Alphandéry, 1994, p. 333).

Germany

In Germany the EC agri-environmental measures starting in 1985 do not mark the beginning of environmental intervention in agriculture and landscape. Whereas nature protection and landscape management have been practised for many years without any debates on agricultural land use and without legal codification of 'good agricultural practice' in ecological terms,[24] agri-environmental programmes have been designed in some federal states since the early 1980s. Such measures (for the reduction of fertilizers and agro-chemicals, for protection of biotopes, certain species and natural resources, mainly groundwater) can be seen as cornerstones of an extensification policy and soon they re-emerged within EU programmes. Without saying that the measures formulated in the 1985 EC-efficiency regulation and later on in regulation 2078/92 are of German authorship, it can be said that they fitted well into the situation and the agricultural policy of the German federal states (although, at first, there was not sufficient support among the farming community). A point to note in several federal states of Germany is that some of the agri-environmental measures now defined as components of regional programmes to implement regulation 2078/92 had quite a career. They were part of regional (federal state) programmes in the 1980s, then became part of the first EU regulation and finally part of the new regulation EC 2078/92 – and, presumably they are waiting for a new EU programme to be continued after 1997.

With regard to the social acceptance and impact of the agri-environmental measures, the longer the measures are in existence and the more they become generalized or offered to all farmers (not only those in sensitive areas) with compensation payments, the more they are accepted by the farming community and their political representatives. Furthermore, the longer they are in existence, the more actors are trying to make sense of them. The environmental community tries to find positive contributions to nature protection, and the hitherto dominant critical argument, rejecting the measures as inconsequential and insufficient, begins to be used in a less fundamental and more discriminating way. The positions of the opposing camps have been simple at first – most environmental movements rejected the measures because of their insufficient contribution to environmental protection and most farmers refused to accept them because the represented state intervention in their production systems. Now the lines of demarcation are beginning to fade away.[25]

Portugal

In Portugal, agri-environmental measures such as the first big environmental programme covering agriculture, have to be related to recent changes in agriculture to be interpreted adequately. Traditional agriculture in Portugal has been extensive, with low levels of nutrient input and high levels of semi-natural vegetation. Although it may cause environmental problems, for example by overgrazing or excessive use of water, with its extensive character it was environmentally sound. Since Portuguese agriculture is still predominantly not intensive, there is not much point to introduce extensification measures targeted at de-intensifying forms of production that are still not typical for the country, practised in limited areas and by a minority of farmers only.[26] The implementation of agri-environmental measures becomes more meaningful when they can be used to reach other goals at national and regional level rather than at EU level. These goals can be understood when some of the dramatic changes within rural Portugal have been analysed – changes during the period in which Portugal came under EU rule. Between 1981 and 1991 the agricultural population declined from 25.7 to 16.7 per cent. And, as a consequence of outmigration from rural areas, more and more rural residents are elderly people. In 1989 one third of the Portuguese agricultural population was over 55 years old. In the interior rural areas demographic recession combined with a low standard of living and accentuated ageing of the population have already contributed to the physical and human abandonment of the countryside. The level of population density in the interior rural areas has fallen dramatically, social and physical desertification becoming a real threat.[27]

The main environmental problems in rural areas are those caused by abandoning land and by rural exodus, but these do not fit into the goal structure of regulation 2078/92, which is aimed at market relief through extensification. To close the gap between real local problems and problems as they are defined by the EU, the measures have to be redefined in such a way that they do not just support extensification, but mainly serve to support the continuation of traditional locally adapted forms of agriculture, and become components of social programmes, helping to reduce migration, because the of rural landscape and environment can only be preserved if the rural exodus is stopped.

The application of agri-environmental measures in Portugal is therefore aimed at combining preservation of the natural environment with redressing some of the particularly problematic social and demographic trends in rural areas. The measures in the Portuguese version of regulation 2078/92 are intended to ameliorate social conditions while maintaining and preserving extensification and traditional systems of agriculture (Instituto de Estruturas Agrárias e Desenvolvimeno Rural, 1994).

European agri-environmental policy and national interests

Deconstructing the explanation and action logic of regulation 2078/92, we may find an ideal-type argument like the following: to reduce over-production in agriculture and to contribute to more environmentally sound forms of using natural resources, all types of measures for reducing intensity of production at farm level – from reduction of inputs like fertilizers and agro-chemicals (as weak forms) to conversion to organic farming (as a strong form) – will be supported by CAP. This argument reveals the preponderance of 'northern problems' and the northern, rich countries in the EU. Furthermore, it is not a social construction in the sense of socially anchored forms and traditions of agriculture, but a specific scientific-bureaucratic construction based on farm management. It is perceived as a correction to the former course of CAP rather than as a reaction to real developments in the agriculture of member states. However, with their influence on the national discourses, the agri-environmental measures of the EU seem to be the first step in a process of building a European policy on rural environment outside the traditional scope of expert circles, bureaucratic policy and the agricultural lobbies, a policy which demands social legitimization in the member countries.

The role of the EU in national discussions of agriculture, environment and rurality

The influence of the EU in the member states is not perceived mainly as a new supra-national or 'originally European' approach to environmental issues, but through a re-definition of the social role and legitimacy of agriculture in the various countries.

In France, EU environmental policy and agri-environmental measures seem to have had an accelerating effect on the unfolding of agri-environmental discourses in the wider society, although the breaking of the taboo happened late (Deverre, 1994). Inside French society the rules of the game have changed for the agricultural community, but agriculture has not yet found a new and stable identity (Billaud, 1994).

In Germany, the new critical discourse on environmental problems of modern agriculture and destruction of agrarian landscapes was not perceived to be of EU origin but to have arisen as issues at national, regional and local level unfolded politically in the controversies between the environmental and agricultural lobby, with the agricultural lobby being and staying the dominant one. The idea of redefining the farmer as a landscape manager and the search for new roles for agriculture is ongoing, but so far these are reactions only to the environmental problems as they arise individually, not an autonomous seeking for a new role.[28]

In Portugal, the agri-environmental measures of regulation 2078/92 as first contact with EU agri-environmental policy (former regulations have not been applied) are not perceived primarily as environmental measures but as part of CAP reform in general. This reform is aiming at reducing over-production and market relief – again problems to which Portugal has not contributed. Thus national agricultural policy and the agricultural community are torn between the aspiration to make use of a policy tailored for the northern countries mainly and the desire for further modernization of Portuguese agriculture which is not able to meet the national demand for agricultural products.

Through its agricultural and environmental policies the EU had a somewhat different impact on rural development in the three countries studied here. It did not simply homogenize development processes by subsuming everything under one coherent, Europe-wide policy and market system, but the EU development schemes for agriculture create turbulence in the social conflicts and discourses on agricultural development and environment. Nowhere are EU impacts simply seen as gains or losses, positive or negative, but they are perceived as parts of conflicting and contradictory developments. 'Europeanization' of policies does not create a new social reality *sui generis* beyond the local, regional and national levels, nor does the EU remain an abstract and artificial political-institutional system which exists only through its adaptation to national and local societies.[29]

Within European societies and within the agricultural population in various EU countries there has been a refusal to accept the idea of agriculture as a polluting activity. With time lags and in reaction to growing environmental awareness, changing consumer styles, critical environmental movements or (confusing cause and effect) as a reaction to EU policies implemented, this awareness and criticism of the ecological consequences of agricultural modernization happened in all countries. The rational basis on which the agricultural population refuses to accept blame for pollution is the fact that the individual farmer is not the individually responsible polluter: he became himself a dependent producer, dependent on the CAP and the attendant politically enforced development.

With regard to environmental policy it has been asked whether the 'Europeanization' of environmental policy has shifted the social perception and construction of the environment as 'being defined in normative, scientific rather than cultural terms' (Buller, Lowe and Flynn, 1993, p. 192). If we go down to the level of the agri-environmental discourse in the member countries, it might be difficult to verify such a trend empirically, although it exists as a general tendency. The question may be partially answered with the argument that there is no socio-cultural tradition or 'cultural reservoir' for the EU, but only for member countries, hence the main legitimation EU policy can find (beyond the scope of power) is scientific, in the sense of science having a universal character.

Conclusions

Contrary to conventional policy research, which looks for an emerging European style of policy and environmental discourse, the issue discussed here is what feedback effects European integration has had on agri-environmental discourse at national level.

Codifying terms of the relationship between rurality and environment, for example, the social dimensions of village–urban, traditional–modern and local–global, represent different, partly contradictory, partly antagonistic, partly complementary or overlapping constructions of spatial order, landscape and nature.

The different time horizons, directions and particular characteristics of agricultural modernization in the three countries studied help to explain the various social constructions of the rural environment. Beyond the differences, the similarities tell us as much about the unfolding of a European discourse. All constructions and the ongoing political discourses up to now have been codified nationally, clearly demarcated by national cultural traditions and national languages. This demonstrates the difficulties of social acceptance and legitimation of a new level of a European political and social order. The existing mechanisms of social integration and legitimization are not fulfilling such a function, and Europeanization is not creating a new supra-national cultural reservoir (beyond such traditional all-European value systems derived from Christianity or western culture). Although such an argument is based on insufficient sources – besides our country studies which are not systematically oriented to this question, we may refer to discussions in the political sciences (Liefferink, Lowe and Mol, 1993) and, to a certain degree, to evidence and plausibility – it might not require much further scientific verification. It relates to a temporary situation which may change in future.

In France the EU brought about partly unexpected changes in the traditional strong and positively valued role of agriculture, which has to find new forms of legitimation. However, the relationship between agricultural practices and environmental quality is not codified in a single and coherent social construction. It is effective at different political and social levels. The agri-environmental measures are not the only reference point for a new discourse, but they are part of a contradictory process of development. It is through a 'bottom up' process, starting at local level, that these measures become part of the construction of a new agricultural legitimacy; that is, they do not become effective through direct confrontation with national discourses, social constructions, identities and ideologies. At local level they are not simply new social norms that are redirecting behaviour of farmers, but they become part of a process of changing the rules of the game, i.e. of redefining the role and legitimacy of agriculture in an open process. The experimental quality, the flexibility and local adaptability of the measures underline this process and negotiation type of effect. The measures can be influenced, changed, reviewed in the further policy process. Thus time becomes the decisive factor.

In Germany rural and agricultural development took an unexpected course in the 1990s with the end of the East–West division and after German unification.[30] Although not a consequence of EU policies, EU becomes now for the first time a critical opponent to German environmental policy, and German approaches to environmental regulation lose significance. This shifting of the balance has an incalculable influence on future agri-environmental discourse. A first effect of this might be seen in the way that the traditional confrontations and conflicts of interest between the agricultural and environmental community begin to melt away and demand a redefinition of the objectives and roles of agriculture for the environment in the years ahead.

As a consequence of their relative autonomy, the federal states are free to formulate their agri-environmental programmes and prioritize or neglect certain types of measures if they do not fit into regional practice of agriculture. Although most measures can be found in most state programmes, state policies are not made coherent by the federal government. Thus, certain measures are practised in certain regions only and payments for farmers' participation in management agreements vary from state to state. The discriminating impacts of the measures for certain types of holdings and certain regions will be a continuing critical issue in the implementation of agri-environmental policy. What this means for the further development of agriculture, with possible consequences of regional segregation of areas with intensive and extensive agriculture etc., cannot be evaluated at the moment.[31]

In Portugal, there was little environmental and agri-environmental policy before its late accession to membership of the EC. This membership *per se* brought about a rapid and accelerated process of exogenous modernization, part of which is 'ecological modernization.' In this conflicting process historically different forms and systems of agriculture interact, with strange effects. EU agri-environmental policy introduces measures into Portuguese agriculture which are targeted at other forms of agriculture, not yet unfolded here. What does extensification mean in an agriculture which is still predominantly extensive and traditional?

In Portugal, the social discourse on rural environment remains divided into national and 'imported' EU debates which do not link very well with the problems experienced by people and the concerns for improving the quality of life in rural areas. Health and sanitation concerns such as running water and regular garbage collection, are still primarily equated with creating basic modern and developed conditions. In general, the social discourse continues to identify modernization of agriculture as a goal, and preservation of the countryside and the rural environment as a traditional heritage which harmoniously combines nature and social space.

Although common social structures of agriculture have been identified – like peasant family farming, professional structures of self-administration and cooperation, privileged roles of agricultural interest organizations – in West European countries (Rieger, 1995), it is not at this level that EU policies influence agricultural development significantly. For the present phase we could say that one effect of EU agri-environmental policy has been a splitting of social discourses and segregation of environmental practices. Such 'reframing' of the concepts and discourses on rurality, agriculture and the environment can be described in the case of Portugal as a contradictory 'dual codification' of rural development. In the case of France, the contradictions become effective at other levels, in many-sided local processes of redefining agricultural identity. In the case of Germany the contradictions may be characterized as being between manifest traditions and manifest developments at socio-economic level.

However, do the agri-environmental measures contribute to a new social legitimization of agriculture in the wider society and does the agri-environmental policy of EU have a catalytic function in this respect?

For France and Germany, with certain differences, it might be said, that a process of social redefinition of agriculture is underway, moving towards a socially legitimized form of agriculture as a new, environmentally sound production activity. The agri-environmental measures of the EU support this tendency. The European influence in both countries seems to be rather similar in respect to the weakening of traditional forms of peasant farming through intensification and at the same time revaluating these forms as a source of environmentally sound practice. In Portugal it seems that the rapid structural change of agriculture dominates the discourse, less a social redefinition of agriculture, and the problems experienced in rural development differ from those of the other two countries. EU policy does not create the new social legitimization – that is still coming from the source of intra-social or national and regional ideologies. Only in broad general trends we might see similarities – for instance in the change from the old legitimacy based on security of food supply to a new one of contributing to the maintenance of landscapes and protection of nature (Deverre, 1994).[32] However, the question whether EU agri-environmental policy is necessary at all, seems a speculative one. This policy is part of a political structure within which it may gradually succeed in redirecting the CAP from productivist to sustainable approaches.

Notes

1. This Chapter is a revised version of the paper presented at the XVIth Congress of the European Society for Rural Sociology in Prague 1995. The authors owe special

thanks to Christof Riegert, Aida Valadas da Lima and Eleuterio Sampaio for their collaboration. The critical comments of Henk de Haan have been very helpful for the revision of the paper, therefore our thanks are also addressed to him. An earlier version of the paper has been published in *Zemedelská Ekonomika*, 42, 1996 (1), pp. 1–10. The research project within which we wrote the paper ended in 1996. A final report entitled 'Sociological Enquiry into the Conditions required for the Success of the Supporting Environmental Measures Within the Reform of the Common Agricultural Policy' is available from the project coordinator, Katalyse Institut für angewandte Umweltforschung (Cologne, Germany).

2. The shortcomings of this new system became visible with its ongoing integration. As a political system it is bureaucratic, highly dependent on changing political majorities and bureaucratic elites and interest coalitions in the member states. It lacks legitimation since there is no European public, there are no European parties and parliamentary decision making is still mainly functioning at national and regional level. As a political system, the EU is neither a state nor simply an international regime.
3. See Liefferink, Lowe and Mol (1993). With the Maastricht Treaty and the completion of the Single Europe Market, enlarged environmental activity of the EU threatens already achieved environmental measures. Centralization happens simultaneously with an effort to decentralize decision making in the EU via the subsidiarity principle.
4. The concept of 'national discourse' is of methodological significance as unit of comparison which makes it possible to demarcate discourses according to a country in the sense of a nation state, with a specific language, history and cultural tradition. The analysis in this Chapter is limited to the purpose of formulating a framework for the empirical analysis of agri-environmental measures of CAP reform, which is the core of our research project. The project came under EU research programme Socio-Economic Environmental Research (SEER) of DG XII. Within this programme similar research has been done for Denmark, the Netherlands and Spain.
5. The choice of countries represents the two dominant models of rural Europe among EU member states until the 1990s – 'a southern model, where the dominant concern is retaining environmental quality while encouraging industrial growth and agricultural modernization' and 'a north-west European form, characterized by land-use conflict and an essentially consumption-oriented conception of nature and the environment' (Hoggart, Buller and Black, 1995, p. 231).
6. The concepts of 'social construction' and 'discourse' indicate the increasing theoretical and methodological influence of constructivist and interactionist approaches in environmental sociology (Hannigan, 1995). The basic ideas of these approaches are clear, expressing a negation of the naturalistic and objectivistic constructions of rurality and environment which have been dominant in the natural sciences but also in the new environmental movements. However, the theoretical elaboration of constructivist thinking unfolds into complicated and heterogeneous versions of sociological theory. Unanimous rejection of the naturalistic fallacy in defining the interaction between society and nature brings together such divergent approaches as the theories of risk society and reflexive modernization (Ulrich Beck, Anthony

Giddens), the theory of autopoietic systems (Niklas Luhmann), the cultural sociology of Klaus Eder and the theory of communicative action of Jürgen Habermas. These are some of important theories influencing environmental sociology. We use the concepts in a generic form (without following a specific theory) as conceptual tools to organize empirical findings and the description of historical developments in the countries studied. The description of social constructions is based on a Europe-wide analysis of the state of the art in social scientific research on rural environmental problems, in which some of the authors of this Chapter took part (Ministère de l'Enseignement Superieur et de la Recherche, Centre National de la Recherche Scientifique, 1994). The concept of discourse relates to a thematically organized unit of analysis not to a specific methodology of analysis.

7. The agri-environmental policy of the EC gained momentum after 1985. It gave rise to extensification programmes: at first the so-called 'efficiency regulation,' later on the accompanying measures of the 1992 CAP reform, formulated in EU regulation 2078/92. These measures have two objectives: market relief and environmental protection, thus becoming ambivalent.
8. The general concepts of city and countryside are multi-dimensional, so conceptual levels have to be specified. For the discussion here the dichotomy is limited to socio-historical spheres of everyday life. More detailed descriptions of traditional social constructions of rurality as they evolved during agricultural modernization from a country comparison can be found in Hoggart, Buller and Black (1995) and Jollivet and Eizner (1996).
9. For the development of French rural society and the role of peasants in the nineteenth century, see Moulin (1988). After the Napoleonic wars until 1870 we find here (and similarly in Germany) at first slow and steady social change with increasing productivity of agriculture. At this period the peasants and rural society were at the peak of their social significance. Towards the end of the century a disastrous fall in prices heralded the first agricultural crisis. Changes in rural traditions and ways of life between 1870 and 1914 add up to an increasing uniformity of rural society which loses its former diversity and becomes more and more identifiable only in relation to urban life, but also comes under increasing influence from urban life.
10. *Ordre éternel des champs* was the title of a well-known book published after the Second World War.
11. For France as for other West European countries it might be said that the consequences of the changes in rural areas became socially important only very recently since the 1980s, together with the 'increasing prioritization of green issues both in social discourse and in political action.' (Hoggart, Buller and Black, 1995, p. 229). This argument is similar to Lutz's (1984) account of the retardation of the social impact of agricultural modernization.
12. In the late eighteenth century in Germany there was a sentimental cultural movement among the aristocratic and bourgeois social strata, which culminated in the celebration of the beauty and simplicity of rural life, inspired by the writings of Rousseau. In the early nineteenth century we find a similar esteem of rurality among

leading intellectuals, in the philosophical and *belles lettres* literature of German classicism and romanticism.

13. Latifundial farms and manor estates as relics of an older social organization of agriculture have existed mainly in East Germany (*ostelbische Junker*), but these have not become the focus of dominant social ideologies in the way that family farms have.
14. See Pongratz (1988). Both concepts remained prominent in theoretical sociology for decades, until the theoretical works of Parsons, although their explanatory quality never was very highly developed.
15. Also in France simple equations like 'environment=nature=landscape' as popular images of urban origin, are discussed (Mathieu and Jollivet, 1989).
16. Traditional denotations of rurality are rooted in the pre-industrial and pre-modern forms of society. The relatively late and rapid modernization of agriculture in both parts of Germany after the Second World War has not yet superseded these social constructions of agriculture and countryside, although socialist development nearly wiped them out. Social constructions of rurality change more slowly than social structures of the society. Moreover, in these social constructions (which are reinforced in the ecological discourse), the pre-modern practice of agriculture survives as a norm of good agricultural practice, and the local level gains new significance for the balanced interaction of nature and society through agriculture. For a discussion of rural development in the framework of cultural ecology, oriented to the analysis of local accommodation of agriculture to the ecological environment (Bruckmeier, 1992).
17. Even in 1950 the active rural population was 50 per cent of total population, and only from the 1960s onwards has migration from the countryside to urban areas steadily increased, providing cheap labour for industrial development.
18. Rural society has been studied from technical, literary and social perspectives in an effort to examine not only the 'idyllic rural life' but also specific problems of agricultural production. Ethnologists, geographers and anthropologists tended to approach the subject of rural society as a closed, self-sufficient entity, as a community. Preserving the oral tradition meant providing interpretations of popular wisdom and knowledge, including its myths, legends and customs. Only since the 1970s have sociology and rural sociology developed in Portugal, and with them the study of rural society expanded (Mansinho and Schmidt, 1994). It was no longer seen only from an internal perspective, including lifestyles and practices, but the rural became situated within the larger historical and political context of the rapidly modernizing Portugal. The 'externality' of rural society presented a new vision, and with it new roles and functions for rural society began to be discussed (Lima, 1987). In both the earlier rural monographs and in the more recent rural sociological studies, rural society has been constructed through a combined analysis of social structures and relations with physical and natural environment. Physical space is therefore not a passive dimension of social relations but part of its very form, and thereby resists, adapts and conditions those social relations.
19. In France the dominant topics, which are also reflected in rural sociology, are agricultural organizations and the future of peasants.

20. Since many big nuclear and infrastructural projects have been built in rural areas, the critical environmental movements are often rooted in local initiatives in rural areas (Rucht, 1993).
21. Years before, since the mid-1970s, a new interest in the application of critical social sciences for rural areas and the crisis of rural life worlds could be observed (Schmals and Voigt, 1986).
22. In a generalized and theoretical form Lutz (1984) has elaborated the thesis that there is no endogenous development pattern for modern capitalist societies in Europe, but that until mid-twentieth century a symbiosis of industrial-capitalist and traditional (peasant and craft-based) segments was prevailing. This symbiosis has changed in the second half of this century, with the consequences gradually becoming visible towards the end of the century. It might be added, that the agri-environmental discourse is part of this manifestation of the consequences of integration of agriculture into globalizing economies.
23. See Alphandéry (1994) for a description of the PDD.
24. The Federal Nature Protection Act from 1976 is subject to alteration since 1996 only. The agricultural clause of this law says that orderly agriculture is contributing to nature protection – this privileged clause for agriculture has long been the object of environmental criticism.
25. See Heinze and Voelzkow (1993) for an analysis of the changing position of the German Farmers' Association (DBV) which follows a strategy of economizing ecological issues, with 'demands for compensatory payments for environmental contributions' of farmers (p. 36).
26. Portugal does not produce an agricultural surplus, its production being insufficient to meet current demands. In the last two decades the situation has got worse, with less than one third of food demand being met by internal production.
27. Other changes in the use of the rural environment have accompanied the decline of agriculture and the active agricultural population: increase of leisure activities such as golf courses, rural tourism, natural reserves and wildlife parks, conversion of farms into weekend houses for wealthy urbanites. However, such changes are slow and their impact has been mostly felt by wealthy tourists and city dwellers, not by rural residents.
28. EC environmental policy and agri-environmental measures before 1990 did not cause a critical disturbance of national and regional environmental policies. German environmental policy reached crisis point after German unification and with the further development of EU environmental policy. Now the national approaches and experiences of environmental policy have become devalued and are coming under legitimation pressure on account of their inefficient, regulatory styles.
29. This applies for the present time when the emerging political and social order between global and local levels still cannot be conceptualized adequately.
30. The reunion of the country with two very different social systems resulted in a new situation. After the end of the political division we find a continuing division of agricultural structures, which has unexpected impacts on West German agriculture,

dominant in Germany up to now. The confrontation of a rather small-scale with a large-scale agriculture will shift the development paradigm from a preponderance of family farming to industrialized agriculture.

31. In most federal states the programmes do not require the participating farmers to bring their total area in a management agreement. Hence they may bring in only their grassland and continue intensive production practices on their arable land.
32. For France, as said above, the point of local differentiation of the measures has to be stressed: local application is the basis for the construction of new agri-environmental legitimacies, not a uniform national-level process.

References

Almeida, J.F. de (1994), 'Sobre a monografia rural,' *Analise Social*, No. 125.

Alphandéry, P. (1994), 'Agricultural Practices and Environmental Perceptions in the Manche Département.' *Sociologia Ruralis*, Vol. 34, No. 4, pp. 329–339.

Bechmann, A. (1988), *Evolutionäre Umweltpolitik – Von der Umweltkrise zur Gesellschaftsveränderung*, TU Berlin, Institut für Landschaftsökonomie: Berlin.

Billaud, J.-P. (1994), 'De la solución negociada de los conflictos a la negociación institucional: la agricultura busca una nueva legitimidad,' *Agricultura y Sociedad*, No. 71, pp. 209–242.

Born, M. (1989) 2nd edition, *Die Entwicklung der deutschen Agrarlandschaft*, Wissenschaftliche Buchgesellschaft: Darmstadt.

Bruckmeier, K. (1992), 'Rurale Gesellschaften aus kulturökologischer Perspektive,' in Glaser, B. and Teherani-Krönner, P. (eds), *Humanökologie und Kulturökologie*, Westdeutscher Verlag: Opladen

Buller, H., Lowe, P. and Flynn, A. (1993), 'National Responses to the Europeanisation of Environmental Policy; A Selective Review of Comparative Research,' in Liefferink, J.D., Lowe, P.D. and Mol, A.P.J. (eds), *European Integration and Environmental Policy,* Belhaven Press: London and New York.

Deverre, C. (1994), 'Rare Birds and Flocks. Agriculture and Social Legitimization of Environmental Protection,' in Symes, D., Jansen, A.J. (eds), *Agricultural Restructuring and Rural Change in Europe*, Department of Sociology, Agricultural University: Wageningen.

Hannigan, J.A. (1995), *Environmental Sociology. A Social Constructivist Perspective*, Routledge: New York.

Heinze, R. and Voelzkow, H. (1993), 'Organizational Problems for the German Farmers' Association and Alternative Policy Options,' *Sociologia Ruralis*, Vol. 33, No. 1, pp. 25–41.

Hoggart, K., Buller, H. and Black, R. (1995), *Rural Europe – Identity and Change*, Arnold: London.

Instituto de Estruturas Agrárias e Desenvolvimeno Rural, Ministério da Agricultura (eds) (1994), *Medidas Agro-Ambientais*, Lisbon.

Jollivet, M. and Eizner, N. (eds) (1996), *L'Europe et ses campagnes*, Presse de Sciences Po: Paris.

Küster, H. (1995), *Geschichte der Landschaft in Mitteleuropa*, C.H. Beck: München.

Liefferink, J.D., Lowe, P.D. and Mol, A.P.J. (eds) (1993), *European Integration and Environmental Policy*, Belhaven: London and New York.

Lima, A.V. (1987), 'Relatório 'Fin des paysans/nouveaux paysans,' *Previsao e Avaliacao em Ciencias e Tecnologia. Programa FAST*, Vol. 1

Lutz, B. (1984), *Der kurze Traum immerwährender Prosperität*, Campus: Frankfurt/Main und New York.

Mansinho, M.I., Schmidt, L. (1994), 'A emergencia do ambiente nas ciencias sociais: analise de um inventaraio bibliografico,' *Analise Social*, No. 125.

Mathieu, N. and Jollivet, M. (eds) (1989), *Du rural à l'environnement. La question de la nature aujourd'hui*, A.R.F. Editions et L'Harmattan: Paris.

Ministère de l'Enseignement Superieur et de la Recherche/Centre National de la Recherche Scientifique (1994), *Bilan des recherches en sciences sociales sur les problèmes d'environnement en milieu rural dans les pays européens*, CNRS-GRMSE, Paris/Nanterre).

Moulin, A. (1988), *Les paysans dans la société francaise de la revolution à nos jours*, Editions du Seuil: Paris.

OECD (1989), *Agricultural and Environmental Policies. Opportunities for Integration*, OECD: Paris.

OECD (1993), *Agricultural and Environmental Policy Integration: Recent Progress and New Directions*, OECD: Paris.

Pongratz, H. (1988), 'Die segmentierte Gesellschaft. Dualistische Gesellschaftskonzeptionen als theoretische Grundlage der Agrarsoziologie,' in Agrarsoziale Gesellschaft (ed.), *Ländliche Gesellschaft im Umbruch. Beiträge zur agrarsoziologischen Diskussion*, No. 101, pp. 54–98.

Rieger, E. (1995), 'Der Wandel der Landwirtschaft in der Europäischen Union,' *Kölner Zeitschrift für Soziologie und Sozialpsychologie*, Vol. 47, No. 1, pp. 65–93.

Rucht, D. (1993), 'Think Globally, Act Locally? Needs, Forms and Problems of Cross-National Cooperation among Environmental Groups,' in Liefferink, J.D., Lowe, P.D. and Mol, A.P.J. (eds), *European Integration and Environmental Policy*, Belhaven Press: London and New York.

Schmals, K.M. and Voigt, R. (eds) (1986), *Krise ländlicher Lebenswelten. Analysen, Erklärungsansätze und Lösungsperspektiven*, Campus: Frankfurt/Main und New York.

Simmons, I.G. (1993), *Interpreting Nature. Cultural Constructions of the Environment*, Routledge: London and New York.

SRU (Sachverständigenrat für Umweltfragen), (1985), *Umweltprobleme der Landwirtschaft. Sondergutachten vom März 1985*, Kohlhammer: Stuttgart und Mainz.

Viard, J. (1990), *Le tiers espace – essai sur la nature*, Méridiens Klincksieck: Paris.

2 Frontiers of consumption: sustainable rural economies and societies in the next century?

Michael Redclift

The theme of this book is 'sustainable rural development.' Let us begin with the goal of sustainable rural development. Late industrial societies have set themselves various environmental targets for achieving 'sustainable development.' In this Chapter I will contrast the ways in which both industrial and 'natural' systems serve to transform the environment. This forces us to consider not only rural economies and societies, but human consumption and the use of natural resources in general.

In the case of natural systems this transformation, or metabolism creates by-products, or nutrients, which become assimilated in the system itself. In this way vegetable (and animal) matter contributes to life itself on the forest floor, in the composition of soils and in the world's oceans. However, industrial systems work differently; nutrients become 'waste' and re-enter the environment in ways that are frequently harmful to ecological systems and to humans themselves.

The question posed by this Chapter links both natural and human-produced systems, and concerns human responsibilities for the environment. It is discussed at greater length in Redclift (1996). To what extent does the transformation of an industrial system highly dependent on 'nature,' like that of agriculture, serve to undermine the sustainability of rural economies and societies? Can such transformations actually enhance sustainability? To answer these questions we need to address some fundamental issues about sustainability itself.

We need to begin by examining what we mean by sustainable levels of resource exploitation within a society. This problem of sustainable levels of resource use strikes at the balance between the production of goods and services and the rate at which we consume resources. Is a 'sustainable' resource system compatible with given levels of consumption, and at what level (local, regional, global)? The technical discussion in the environmental policy literature includes concepts like critical loads, carrying capacity, and ways of assessing biodiversity losses, and non-reversible environmental costs (critical natural capital). . .

The much less obvious issue, which scarcely plays any part in most of the literature debate, is not about what constitutes 'sustainable levels,' but about how they might be *achieved*. In this area most suggestions point towards global environmental management (*pace* 1992 United Nations Conference on Environment and Development – UNCED) usually linked to local action (Agenda 21). As we shall see, arriving at 'rational' grounds for environmental action has proved difficult in practice.

It is clear that we cannot arrive at sustainable levels of resource exploitation without some notion of the way in which resources should be valued. It follows that we need to establish how societies arrive at ways of valuing environmental losses and gains. This is a task in which rural sociologists have much to offer the rather limited concept of 'value' employed by most neo-classical economists. The question usually posed by most mainstream environmental economists is this: how much are people willing to pay for the environment?[1]

Again, the formal question obscures a much more important primary issue: do valuation criteria reflect real social choices, and according to whose understanding? The ways we arrive at environmental valuations can take several forms, including mechanisms to try and achieve social consensus and (more commonly) the conclusions of 'expert' witnesses. It is notable that the Green agenda on decision making has been advanced almost entirely by groups outside conventional political parties. At the same time, mainstream political discourse has not addressed sustainability.

Finally, sustainability does not merely depend upon what is to be sustained, and the value of its conservation. There is also a third issue, which strikes at the relationship between members of a society and the economic and social behaviour that is expected of them. This is: how do societies influence the social and economic behaviour of their citizens towards more concern with sustainability? How do different societies *construct* their views of the environment?

The problems identified above, that questions are posed in ways which seek to *depoliticize the debate*, are even more evident in this case. Most of societies' institutions, from the most formal (government, law, church) to the least formal (family, work, community), take no conscious account of 'sustainability.' Can we steal it in, as it were, (as most commentators hope) or do we have to begin again? Where does good example begin, in the North, which has shown very little commitment so far to reduced consumption, or in the South, where most people consume to survive?

Clearly we should begin to consider the institutions that we bequeath future generations, rather than merely the environment we hand on to them. However, the fact that the question is posed in this way at all, that sustainability is increasingly viewed as an issue alongside governance, tells us something about the shortage of ready-to-wear clothes in the economic policy closet. Unlike the post-War Keynesian medicine, the remedy this time around looks more challenging, and threatens to require some fundamental shifts in our 'getting and spending . . .,' in that ubiquitous shibboleth, '. . . our lifestyles.' Rather than being asked to spend more (Keynes), Green economics asks us to spend less.

In the next section I want to consider the way these questions are posed and answered from a specific point of departure: that of the modern food system. The production and consumption of food, within the modern food system, has developed in specific ways that carry implications for the production process itself and, as we shall see, the environment in which food is both produced and consumed. This rural environment is the object of as much attention in Europe today as agriculture itself.

The dynamics of production and consumption processes within the food system

In most societies there is an ambiguous relationship between the volume of output and the extent of environmental destruction. High output provides resources for environmental protection, but the benefits of conserving resources may be outweighed by the environmental impact of production and consumption itself. On the other hand, although low output provides fewer resources for environmental protection, it may be grossly inefficient and do little to prevent environmental degradation. In the 1970s the principal fear in environmental policy circles was the supposed 'limits to growth' (Meadows, 1972).

In the 1990s 'limits to sink capacities' (or output) poses an equivalent fear, particularly in the context of global problems such as the destruction of the ozone layer and the enhanced greenhouse effect. However, we should not forget that 'natural' carbon sinks, such as forests, grasslands and water courses, are also the location of most rural peoples' livelihood activities. They are not merely part of a throughput of materials and energy, but the location of rural societies and cultures.

The complex interactions between economic activities and the natural environment depend crucially on the way a society organizes its relationship to the environment, and the view that different social groups take of this relationship. Many environmental changes are represented as 'demand driven,' in the hands of consumers, rather than 'supply driven,' in the hands of the formal economic levers dictating production. In fact it is almost impossible to separate patterns of consumption and 'lifestyles' from economic instruments and ideologies. Most production in late industrial societies is geared to increasing volumes, rather than the life-cycle effects of goods and services.

The existence of new materials and productive processes has accelerated the changes through which environmental costs are transported in space (and frequently in time) usually to poorer developing countries. At the same time, levels of air pollution, for example, in the newly developing countries of East Asia, are increasing more rapidly than their increase in Gross Domestic Product. The social processes at work are not simply in the hands of consumers. They are embodied in the global political economy of market capitalism.

Some industries in the North, aware of the costs of employing dirty technologies on their doorstep, have sought either to export their pollution and wastes or to internalize the problems associated with unmanageable levels of waste. They have turned their attention to changing material flows and waste streams as a focus for technological innovation itself.

The concept of eco-efficiency is a response to this crisis. Some companies are learning how to maintain or expand output, while at the same time cutting resource inputs and environmental impacts to a minimum. In similar fashion the same principles can be applied to households, rather than companies. One might then examine the use that households could make of resources to improve welfare without increasing aggregate levels of personal consumption to the point at which they are not sustainable.

Examples are the way in which reductions in product size can facilitate reductions in the time taken to transport goods and services, thereby maximizing the efficiency in the use of space and time. Combining the quality of output and environmental protection in single low-impact technologies, such as combined heat and power generation for energy utilities, is another example.

Similarly, the agriculture and food sectors provide many illustrations of the huge scope for policies which make a realistic assessment of environmental impacts and benefits, replacing the 'value added' in the industrialization of food (packaging, expensive inputs etc.) by benefits from local provisioning, reduced packaging and more attention to the nutritional quality of food. However, such changes require leverage at several points in a society: from urban consumers who demand safer and healthier food, from citizens who want to reduce waste, and from farmers and food retailers who seek to provide local markets.

Such changes in the way products and services are valued also bring in their train significant shifts in the use of space and the allocation of time. Planning for such developments requires a radical overhaul not so much of consumer attitudes as the whole infrastructure of modern living. This applies equally to both the rural and urban spheres. It is impossible to pose a serious challenge to modern food production from within the rural sphere alone.

Exploring the dynamics of consumption and production processes takes us well beyond environmental protection. It affords the possibility, in principle, for societies to use the smaller stock of resources available in a more efficient way. At the moment environmental management is principally concerned with modifying existing human behaviour. Policies are concerned with reducing the full environmental impact of our actions. The underlying behaviour, or social commitments, associated with lifestyles, is currently viewed as non-negotiable. I would like to argue that developing sustainable rural economies and societies will essentially mean opening up the food system and the environment in rural areas to political negotiation. Sustainable rural economies need to be built on the preferences and alliances of groups of people in the wider society. They cannot be constructed from within rural economies alone, or the specialized sector of European economies that is responsible for only a fraction of total consumption and only a small minority of the employed population.

We also need to look outside Europe, in the narrow sense, to the developing countries. In a paper written for the World Bank, Robert Goodland has argued that achieving *per capita* income levels in low income countries of $1,500 to $2,000 (rather than OECD's $21,000 average) is quite possible. This would represent a significant advance in welfare for such countries. 'Moreover,' he adds, 'that level of income may provide 80 per cent of the basic welfare provided by a $20,000 income, as measured by life expectancy, nutrition, and education' (Goodland, 1994). There might also be measurable gains in personal security and reduced social tensions in societies experiencing this type of transition. Goodland then suggests that 'Colleagues working on Northern overconsumption should address the corollary. Can $21,000 *per capita* countries cut their consumption by a factor of ten and suffer "only" a 20 per cent loss of basic welfare' (Goodland, 1994). This argument illustrates the way in which thinking about sustainability forces us to think also about the constituents of social welfare and the political conditions under which gains in both sustainability and welfare might be made.

If underlying commitments are non-negotiable then there is little point in pressing on with this type of analysis. However, the problem of overconsumption might be approached from the other direction. Given that certain resource limits are binding, how can we live within them, producing and consuming goods and services in a way that carries net benefits for society in the future? Arguably, a better understanding of production and consumption flows could help societies make radical choices at different levels of analysis and on different time scales, which transferred welfare benefits between countries, as well as within them, making 'development' a much more reflexive and redistributive process than it is today. For those of us who are rural sociologists, it means exploring the meaning of 'welfare' in the societies which we know best, and the means of satisfying the needs of people in these societies – existential needs as well as material ones.

At this stage it is worth pausing to consider some of the difficulties with this line of argument:

1. The view of sustainability I have outlined suggests that many environmental problems, such as the management of wastes and pollution, arise out of the inherent differences between 'natural' and 'industrial' systems. The laws of thermodynamics have precedence over those of the market. At a philosophical level, this may be true. However, efforts to quantify the life-cycle costs of the production and use of goods and services are difficult to express in alternative economic terms. It is not easy to arrive at the utilization values which correspond with actual environmental costs and to substitute them for current market values. Without such utilization values it is impossible to alter the economic behaviour of individuals, still less that of governments. But, can utilization values be derived independently of the 'actually existing' economic behaviour of individuals? And how would we undertake valuations for the products we eat and drink without incorporating 'externalities,' such as the effect of their production, on our view of 'nature' and the condition of the physical environment?

2. Being able to arrive at sustainable levels of exploitation in *environmental* terms – conserving resources and enhancing sinks – tells us nothing about the social and economic mechanisms that will be required to achieve these objectives. Nor does it tell us about the processes through which sustainability assumes legitimacy with the public. These factors – the social mechanisms to achieve greater sustainability and the legitimacy of sustainability as a social goal – lie outside the provenance of industrial metabolism as understood in the literature. They are linked to wider questions of engagement in the society. We need to turn to other approaches to understand them better. Approaches which have their origin in urban as well as rural societies.

Ways of evaluating environmental gains and losses

As we have seen, the way that values are attached to the environment has been dominated by economists. The point of departure for most economists, particularly those who espouse a neo-classical position, is the trade-off between economic growth and environmental protection. Policies to correct environmental problems, it is argued, inevitably carry costs for economic growth and with it the level of consumption. As Jacobs and Ekins (1995) point out, this concern with the cost of environmental measures serves to disguise the fact that neo-classical economics has considerable difficulty in acknowledging that distributional issues, within and between generations, lie at the heart of valuation. The 'willingness to pay' axiom, with which environmental goods are accorded market values, sets aside the central issues which beset the policy agenda: who should pay, and when?

The environmental economics literature focuses upon two related questions which assume importance in a modified neo-classical position: externalities and public goods (Bhaskar and Glyn, 1995). Externalities are important because of the lack of a market in environmental goods. Two solutions are usually proposed. Neo-classical theory has developed around ways of imputing market values to environmental costs and benefits through instruments like contingency valuation. The second solution is to 'internalize' externalities, what has come to be known, especially in Germany and the Netherlands, as *ecological modernization* (Mol, 1994). Quoting Joseph Huber, one of the principal exponents of ecological modernization, Mol refers to this process as '. . . an ecological switch-over of the industrialization process . . . all ways *out of* the environmental crisis lead us further *into industrial society*' (Mol, 1994, p. 11).

Both 'solutions' have problems attached to them. The modified neo-classical position, which imputes market values to nature, starts from a number of assumptions which can be challenged. First, it assumes that individuals act alone to calculate their advantage from making choices under market conditions. The 'individual rational calculator' approach, which lies at the heart of the neo-classical paradigm, provides an approach to decision making that reduces human agency to price signals. There is no place for society in this view of the economy, as several writers have observed (Jacobs, 1991; Benton and Redclift, 1994).

Second, this perspective turns on the relationship between prices and values. We are always in danger, as Oscar Wilde put it, of knowing the price of everything and the value of nothing. We know, for example, that rural landscapes and the countryside have different value for different groups of people, and for some of these groups it is not adequate to translate these as 'intrinsic values' or 'existence values,' for they lie in the abandonment of the market, not its extension.

There are similar problems with ecological modernization. Externalities are not merely environmental costs which can be refashioned into an environment-friendly good or service. They frequently have distributive consequences and causes, which carry political implications for global markets. It is state power, and that of transnational corporations, which frequently lies behind externalities, determining who bears the costs of pollution, toxic wastes, or the effects of pesticide use. Indeed, as Murray (1995) has suggested in a highly original treatise on pesticide use in Latin America, the behaviour of companies which modify agro-chemicals serves to confirm, rather than alter the essential relationships of power which drive the diffusion of modern, industrialized agribusiness. There are limits in the extent to which 'modernization' incorporates ecological considerations.

Finally, it can be observed that some goals of neo-liberal economic policy, such as the liberalization of trade, actually serve to increase externalities. There are basic contradictions between economic growth in the late twentieth century, and the protection and conservation of the natural environment in rural areas. Ecological modernization does not alter these facts.

The paradox at the heart of the economic valuation of the environment is that of distribution. The environment can only be properly 'valued' by successive generations of people, for whom any one generation acts as stewards. Intergenerational equity, moreover, assumes more importance as problems assume 'global' importance, and are increasingly governed by uncertainty. The driving forces behind issues like global warming, the destruction of the ozone layer, and the loss of biodiversity, make little sense within the neo-classical model. (Which is why so much effort is invested in demonstrating that they do not represent a problem for the model). Since unborn generations have no 'rights' it is difficult to rely on efficiency in allocating resources between future and present generations.

The problem of intergenerational equity is matched by that of intra-generational equity. Does every member of the human race count equally in their responsibility for carbon emissions? Do emissions of methane from paddy fields or livestock in poor rural societies count equally with carbon emissions from vehicles being driven between home and shopping mall in California? The most revealing aspects of carbon budgets for nations are that they reveal such vast inequalities in consumption and such different trajectories for future development, that only a complete overhaul of the global economic order would enable any meaningful and lasting international agreements to be made. Intra-generational inequities are currently such that we might as well abandon any pretence that development can be sustainable.

Finally, the problem with the neo-classical valuation of the environment is that it makes questionable assumptions about human behaviour. As a consequence it emphasizes 'efficiency gains' for which the confidence limits are low, over vitally important distributional consequences, for which the confidence limits are high. It succeeds in turning the world on its head. Environmental economics, at least in its mainstream neo-classical version, requires that we ignore the institutional context for decision making, which itself determines whether economic models are used at all.

Social behaviour and sustainability

To meet even modest environmental goals requires significant changes in human behaviour. But how would such changes in behaviour be brought about? If differences in behaviour are based on different perceptions of risk or uncertainty, then it is necessary to begin by explaining the gap between perceptions and behaviour. We all possess enough anecdotal evidence that perceptions influence behaviour in ways that are frequently perverse. The British MET Office (Meteorological Office) has established, for example, that people turn on their central heating in response to cloudy, rather than cold, weather. Can people be relied upon to do what 'rational' action requires, to avert environmental disasters?

This depends on what is meant by 'rational' action; the question at the heart of the debate about so-called risk society and modernity (Beck, 1994). According to Ulrich Beck late industrial societies have removed technology from the political arena. In the process such societies make it more difficult for individuals to deal with risk in everyday life. In Beck's view social problems are viewed, in late industrial societies, as problems of *individual failure*, rather than social competence. When people challenge their society's competence, in particular that of its 'experts' and political leaders, it becomes a challenge to public science. Human agency becomes embroiled in challenges to the authority of science.

The contrasts with mainstream neo-classical economics could not be greater. While economics has been pushing environmental assessment *towards* market-based, individualized models of human behaviour, sociological theory has been emphasizing that environmental risks increasingly reflect individualization and *socially constructed differences*. Economics has sought to bring the environment into the ambit of human decision making by allocating environmental values like any other market value. Sociology is arguing that environmental decision making actually reflects social concerns and commitments, that the environment is nothing more, nor less, than the battleground on which political interests are contested. Social theory is increasingly concerned with the problems that the environment poses *for individuals, rather than by them*.

The differences do not end there. While economics has difficulties, as we have seen, in incorporating distributional issues (time, space and class) into its methodological canon, sociology experiences difficulties in distinguishing between 'what is happening in actually-existing societies' (realist concerns) and the part played by human societies and consciousness in *interpreting* what is happening (constructivism). Some mainstream sociological writing gives the impression that ecological issues can be confined to the margins of society, separated from the social institutions of modern life (the family, work, religion and politics). It is as if our getting and spending were somehow divorced from its social consequences. This is equally true of rural societies. Rural economies and societies in the twenty-first century will not be appendages of urban life, but a stage upon which competing calls on resources are negotiated. Meeting conflicting objectives between nature and society in the rural Europe of the future will not be easy. It is likely to come about as a result of social conflicts between groups of consumers and producers, whose principal aim is not the preservation of a rural idyll or a harmonious past. Rural economies and societies are likely to become a defining aspect of late twentieth-century individualism, and as such an outcome of the wider battlefields of what Ulrich Beck has termed the 'risk society.'

Beck (1994) argues that modernism was founded on ways of exploiting nature and countering tradition. By contrast, post modernism is concerned with the problems of modernism itself, it is '. . . becoming its own theme' (Rustin, 1994, p. 19). Beck grounds his approach, as Rustin avers, in the modernist vision, rather than its sequel. He believes that rational consensus is attainable once the conditions of democratic decision making have been achieved in civil society. Contrary to many, who see human society as constituting a problem for ecological goals (deep ecologists, many eco-feminists), Beck envisages a society which takes responsibility for its own development, as a real, historical possibility.

For Beck contemporary society, in the industrialized world at least, is an incomplete rationality. What we have described as industrial ecology, the search for utilization values which adequately reflect environmental costs, is hampered by the hegemony of the market. The rationality being imposed is the 'wrong rationality,' since, if sustainability is the goal, it forces us to obey the rules of industrialism, rather than those of nature.

Similarly, neo-classical economics finds rationality in its own (implicit) model of human behaviour, which facilitates the efficient allocation of resources. For post-industrial social theory it is rationality itself which is the problem. In Beck's view the *categories* of industrial society prevent the achievement of rational modernity through the three domains of gender, class and ecology. He identifies institutional spheres in which our commitment to the *principles* of industrialism has prevented us from responding adequately to its problems and contradictions. In marxist terms, in modern society much of the superstructure is actively out of engagement with the trajectory of the material base.

Having opened up important questions, *risk society* does not necessarily answer them. The 'individualized social inequality,' which Beck sees as having replaced class, has implications for the relations between labour and the environment, which he fails to develop. He discusses the way individual rights have been invoked in the 1980s *against* forms of social provision. However, modern consumer society raises other, equally important questions. To what extent does the achievement of higher levels of social provision and collective consumption also endanger environmental goals? And, to what extent are environmental goals increasingly a means of expressing political and social goals?[2]

One of the most important, but still insufficiently explored, factors at work in modern society is that of 'ecological risk.' As rural sociologists it is important that we give some attention not only to the risks within rural environments and communities, but also with the way they are perceived by individuals and represented in political terms. Beck sees '. . . solidarity motivated by anxiety . . . (replacing) solidarity motivated by need,' but it is not always clear whether society *should* be more anxious about ecological risks, or that it *is* more anxious. Similarly, the concept of 'need' is itself an interesting candidate for reflexive analysis, since, as I have argued elsewhere, it is unlikely that 'needs' can ever be fully separated from 'wants' (Redclift, 1993).

Nevertheless, Beck's argument, and that of others (Dobson, Merchant, Benton and Redclift) further undermines the separation between 'nature' and 'society' which many of us have been uncomfortable with for some time. As the conquest of nature becomes absolute, eloquently manifested in the modern food system; it is a powerful corrective to conventional class politics. Increasingly social movements can be expected to contest the dominance of nature rather than to further it, although the political expression of these tendencies will vary over time and space. For some it will lie in low-input agriculture, for others in new evocations of rural 'community' (so-called 'neo-rurals'). For others again it may lie in campaigns around animal rights, or the protection of rural livelihoods, as vineyards are replaced by scorched earth.

As science begets more science, and through environmental regulation seeks to 'manage' the consequences of pollution and environmental risk, so civil society comes to represent social and political barriers in environmental terms. Political ecology has come of age.

Of course, alternative lifestyles are still confined to relatively small groups of people in rural Europe. Living in horse-drawn caravans cannot be expected to appeal to most people in rural societies in Europe today. However, if 'lived' alternatives to unsustainable societies are confined to only a few travellers and protestors, nevertheless the principles underlying their protests are no longer as remote from everyday life as they may once have seemed. Increasingly, urban people look to the countryside not only to produce their food, but to provide them with holiday homes and recreational space. What they find may represent a confirmation of industrial society, rather than an antidote to it. The search for an end to alienating work, resistance to valuing nature purely in market terms, the disempowerment associated with traditional occupational roles, are all facets of modern life in rural societies. The frontiers of consumption today do not end at the doors of rural economies and societies, as they so often did in the past. Indeed, answers to fundamental questions about sustainability may even begin there . . .

Notes

1. The exceptions, such as Daly, Norgaard, Ekins and Jacobs are interesting because they take issue, to varying degrees, with the assumptions of neo-classical economics.
2. These issues, in a curiously coded form, are part of the current policy discourse around employment and sustainability in the European Union, encapsulated in the recent White Paper of the European Commission (1993) *Growth, competitiveness and employment* (Fleming, 1994).

References

Beck, U. (1992), *Risk Society: Towards a New Modernity*, Sage: London.

Bhaskar, V. and Glyn, A. (eds) (1995), *The North, the South and the Environment*, United Nations University Press/Earthscan: London.

Daly, H. (1992), *Steady-State Economics*, 2nd Edition, Earthscan: London.

Dobson, A. (1990), *Green Political Thought*, Unwin Hyman: London.

Ekins, P., Hillman, M. and Hutchinson, R. (1992), *Wealth Beyond Measure. An Atlas of New Economics*, Gaia Books: London.

European Commission (1993), *Growth, Competitiveness and Employment*, Brussels.

Fleming, D. (1994), 'Towards the Low-Output Economy: The Future that the Delors White Paper Tries not to Face,' *European Environment*.

Goodland, R. (1994), 'Environmental Sustainability' Working Paper, World Bank: Washington DC.

Intergovernmental Panel on Climate Change (IPCC), Working Group Three (1995), *Second Assessment*, IPCC: London.

Jacobs, M. (1991), *The Green Economy*, Pluto Press: London.

Jacobs, M. and Ekins, P. (1995), 'Environmental Sustainability and the Growth of GDP,' in Bhaskar, V. and Glyn, A. (eds), *The North, the South and the Environment*, United Nations University Press/Earthscan: London.

Meadows, D.C., Randers, J. and Behvene, W. (1972), *The Limits to Growth*, Pan: London.
Merchant, C. (1992), *Radical Ecology*, Routledge: New York.
Mol, A. (1994), 'Ecological Modernization and Institutional Reflexivity,' paper presented at the XIII ISA Congress, Bielefeld.
Murray, D.L. (1994), *Cultivating Crisis: The Human Costs of Pesticide Use in Latin America*, University of Texas: Austin.
Norgaard, R. (1994), *Development Betrayed*, Routledge: London.
Redclift, M.R. (1993), 'Values, Needs, Rights: Reassessing Sustainable Development,' *Environmental Values* Vol. 2, No. 1, pp. 3–20.
Rustin, M. (1994), 'Incomplete Modernity: Ulrich Beck's Risk Society,' *Radical Philosophy*, Vol. 67, pp. 3–12.
Weizsacher, E.U. von (1994), *Earth Politics*, Zed Press: London.
Weale, A. (1992), *The New Politics of Pollution*, Manchester University Press: Manchester.

Part II

Globalization and the Food System

3 Organic food exporting in New Zealand: sustainable agriculture, corporate agribusiness and globalizing food networks[1]

Hugh Campbell

Introduction

Since 1990, organic food has become a small, but growing, contributor to New Zealand's agricultural export earnings. By 1996, the exporting of organic fruit and vegetables has formed an emerging food network, with an estimated value of between NZ$12–15 million. This network has only emerged after considerable contestation at key points, and has involved the development of alliances between consumers, growers and corporate entities. Two aspects of the network mark this development as being unusual. First, organic food production has been strongly biased towards export activity with less than 16 per cent of organic food produced in New Zealand being consumed locally (down from over 90 per cent in 1990). The strong orientation towards export activity is unusual compared to the experience of organic production in other countries. Second, this development has involved an often uneasy relationship between systems of sustainable agriculture and large agribusiness corporations.

The development of organic exporting in New Zealand raises a series of interesting issues. First, what is the role of organic food within the current world food system? Second, what kind of relationship is emerging between large agribusiness corporations and the sustainable agriculture movement? The evidence from New Zealand indicates that organic food is emerging as a distinct feature within the current world food system, and it also indicates that agribusiness corporations are playing a significant part in this development. Such a finding has important implications for the way in which the future of sustainable agriculture is theorized, particularly in the light of recent theories of the world food system.

This Chapter will address these issues by using a case study from New Zealand to link two bodies of literature; the recent literature on world food systems, and the literature theorizing the relationship between sustainable agriculture and agri-food corporations. First, recent theories of change within the world food system will be presented; concentrating on those aspects most relevant to the development of organic food exporting. The theoretical notion of the food network will be presented as the most useful theoretical device for understanding the restructuring of food systems. Second, the case study of organic exporting from New Zealand will be presented and discussed within the framework of food network theory. Finally, attempts to theorize the emerging relationship between agribusiness and sustainable agriculture will be briefly reviewed and critiqued using the evidence from New Zealand. This Chapter will conclude by arguing that food network theory provides a better framework for understanding changes in organic food production and exporting than the prior theoretical focus on corporate entities that was implicit within the sustainable agriculture literature.

Theorizing the world food system

There is a growing body of literature attempting to theorize the contemporary nature of the world food system. Much of the impetus for these attempts was provided by the work of Friedland (1980, 1984) and Friedland et al. (1981). Friedland's advocacy of the commodity systems approach, as exemplified by his study of the Californian lettuce industry, was crucial in directing attention away from a structural analysis of capitalism and family farming and towards an understanding of the historically contingent restructuring of food industries and agro-commodity systems (Buttel et al., 1990).

A second, more recent, theory of the world food system has been articulated by Friedmann and McMichael (1989) and Friedmann (1990, 1991, 1993). Friedmann and McMichael's notion of the food regime combined an understanding of the world food system with the historical periodization of Fordism and post-Fordism being propounded by the French 'regulation school' (see also Kenney et al., 1989, 1991). Central to Friedmann and McMichael's (1989) argument is the notion of the Second Food Regime which is coincidental with the period of Fordist expansion from 1945–73. The Second Food Regime was built around American hegemony in the world food system, promulgating a few key 'complexes' of 'durable' foods. In particular, grains, sugar, fats and oils became the central durable commodities in the establishment of global agriculture along the lines of the post-war structure of American agriculture (Friedmann, 1990, 1991, 1993).

While there have been some significant criticisms of the idea of food regimes and the wider concepts of Fordism and post-Fordism (eg., Goodman and Watts, 1994; Fine et al., 1996), theories of the world food system have proliferated, with a number of important strands emerging in the debate. One of these strands has particular relevance for the study of organic food production in New Zealand.

Commodity systems and globalization

The study of organic food exporting can be usefully explored by using the work of those theorists who have tried to combine the commodity systems tradition with recent notions of globalization.[2] In New Zealand alone there have been numerous attempts to detail changes in different agricultural commodity systems, including: the pastoral sheep/beef commodity system (Le Heron, 1989a, 1989b; Le Heron et al., 1989); citrus production (Le Heron et al., 1992; Le Heron and Pawson, 1996); dairy (Le Heron and Pawson, 1996); and vegetables (Le Heron and Roche, 1995).

Within the international literature, some contributors have attempted to merge the insights of globalization theory with both commodity systems analysis and the regulationist/food regimes genre (Le Heron, 1993; Le Heron and Roche, 1995). This latter, synthetic, group have generally attempted to articulate the transition from the Second Food Regime to some form of successor regime. McMichael (1993) and McMichael and Myhre (1991) have outlined this transition in mainly political terms. In contrast to this, other contributors have attempted to sketch out the successor regime as being constructed around particular food commodities. Friedland (1994) describes the rise in 'fresh' foods as being symptomatic of recent developments in the world food system. These foods are best exemplified by a rise in counter-seasonal fruit, speciality fruit (see also Cook, 1994), and 'fresh' vegetables. Le Heron (1993) and Le Heron and Roche (1995) identify a similar rise in 'fresh' foods and link this to an increasing political sensitivity to issues surrounding 'sustainability' and the initiatives of globalizing capital to operate within (and around) such regulatory structures. Watts (1996) incorporates an even wider range of foods in arguing that fresh fruits and vegetables are only part of the emerging regime of High Value Foods (HVFs) which also includes poultry, dairy and shellfish.

There are distinct similarities between all these attempts to identify the successor commodity systems to those central to the Fordist period. Friedland (1994) suggests, however, that such developments are not likely to provide a total new regime, but in fact are symptomatic of a rising dual market for foods, with the mass-produced and consumed industrial foods persisting for the poorer population while the emerging service and professional classes establish a growing market for fresh or high-value foods. All these contributors are convinced that the market for high-value and/or fresh foods is increasing, and it is towards this market that organic food is targeted. Friedland (1994), Arce and Marsden (1993), Marsden and Arce (1995), Le Heron and Roche (1995), Lawrence (1996) and Watts (1996) all specifically mention organic food production and locate organic food production at the centre of developments in the world food system.[3]

Organic food in the world food system

Friedland (1994, p. 226) argues that even though organic food poses problems of consistency in production, distribution and demand (all counter to prevailing tendencies in the world food system), the rising level of consumer concern over food safety is driving a long-term trend of rising organic sales in the United States. A similar dynamic of consumer concern and the phenomenon of 'food scares' is observed by James (1993) in the UK. Again, organic food consumption is directly linked to such consumer concerns. Arce and Marsden (1993) and Lawrence (1996) put forward a similar position, and locate organic food directly in with the wider consumer move towards fresh and exotic produce. In a more brief assessment of organic food, Watts (1996) speculates that organic food production represents an expression of the 'politics of collective consumption,' but also suggests that the impact of any greening of the food system may be exaggerated. Marsden and Arce (1995) also place organic production in a similar context, namely, the influence of shared ideas and beliefs over healthy and 'ideologically sound' foods which can form 'new consumption niches' linking certain producers, retailers and consumers at a transnational scale. Again, such niches are seen to be strongly indicative of new food relations, while remained fairly marginal in the overall scene.

Within the New Zealand literature, Le Heron and Roche (1995, 1996) briefly assess the emergence of organic exporting in New Zealand. They argue that the development of the Wattie Frozen Foods Ltd. (WFF) organic vegetable exporting programme is symptomatic of the way in which transnational capital is operating in the emerging regime of 'fresh' foods. Consequently, they claim that developments in New Zealand are directly related to wider trends in the world food system.

The theoretical approach used in this Chapter was developed by Campbell (1996) in response to the kinds of theoretical insight outlined above. In particular, Campbell (1996) highlights the contribution of those food theorists who are trying to introduce subjectivity and contestability (Arce and Marsden, 1993; Marsden and Arce, 1995; Marsden, 1996), agricultural and food discourses (James, 1993; Jarosz, 1996), and food symbolism (Cook, 1994) into the food systems literature. Campbell (1996) takes the notion of the food network, as articulated by Marsden and Arce (1995) and applies it to the specific case of organic food exporting in New Zealand. The food network concept involves a recognition of the subjective aspects impacting on food systems at a multitude of levels. It also allows for the study of commodity/food systems without resorting to any form of functionalist or structurally determined analysis. Food networks form according to the historically contingent combination of different economic, social and cultural interests. Most importantly, these interests are contested and combined in different parts of the overall food system.

The following discussion will briefly review the development of organic exporting in New Zealand and then examine the dynamics of this development according to some of the concepts outlined in the theory of food networks.

The emergence of organic food exporting in New Zealand

The following discussion involves data collected as part of a three year programme of research into the developing organic agriculture industry in New Zealand. This programme examined the phenomenon of organic exporting which has experienced a minor boom from 1990–96 and shows few signs of abating. In 1990, a Ministry of Agriculture and Fisheries report (MAF, 1991) estimated that the total value of organic production in New Zealand (90 per cent of which was directed at the domestic market) was around NZ$1.1 million. Since that time there has been a massive increase in the value of organic production, and in the 1995–96 year, the newly formed Organic Products Exporting Group assessed the export value of its members at NZ$12 million. Given that this estimate ignored those few organizations outside the group and also did not assess the size of the domestic market, it could be conservatively estimated that the total annual value of organic product in New Zealand was NZ$15 million by 1996. While this figure is not large by comparison to some of New Zealand's established agricultural export industries it is still around 25 per cent of the export value of wine – one of the most high profile new primary production sectors in New Zealand.

The development of organic exporting has taken place in five regions – Canterbury, Bay of Plenty, Gisborne, Hawkes Bay and Manawatu. Within these regions, two business organizations – the Kiwifruit Marketing Board, and Wattie Frozen Foods Ltd. – have played a prominent part. Both organizations are responding to a significant increase in world trade in organic products. Friedland (1994) documented that organic sales in the United States rose from US$175 million in 1980 to US$1.5 billion in 1990. Recent forecasts suggest that the US market will have reached US$4.4 billion by 1997 (McErlich, 1994). In Canterbury, Wattie Frozen Foods Ltd (WFF) have established a growing export trade in organic vegetables (principally to Japan) while in Bay of Plenty the Kiwifruit Marketing Board is sourcing increasingly large supplies of organic kiwifruit. This Chapter focuses on one of these regions – Canterbury – and investigates the role of WFF in developing an export trade in organic vegetables.

Canterbury is the only major arable producing province in New Zealand as it is one of the few parts that has flat terrain and a dry climate suitable for finishing crops. Historically, farms in Canterbury have tended to mix pastoral and arable activities, with farm businesses swinging between mainly sheep or crop production depending on world prices. This tendency for mixed farming systems combined with a physical environment that is relatively free of pests and diseases (remembering that New Zealand has no close geographical neighbours) means that Canterbury is considered to be very suitable for organic production on mixed farms. Coupled with this is the perception that mixed farmers in Canterbury are very skilled and have recently adopted a wide range of high-value niche products. These products are often produced through contract relations with a number of entrepreneurial agribusiness companies (Campbell, 1994).

Prior to 1990, Wattie Frozen Food Ltd (WFF), a division of the Australasian-based transnational corporation Goodman Fielder Wattie, had established long-term grower relations in Canterbury. WFF sourced a range of non-organic vegetables from Canterbury and these were produced, on contract, in broadacre rotations by mixed pastoral/arable farmers in the region. In 1990–91, WFF contracted its first producer to supply organic vegetables, and the first crop of organic peas was harvested in February 1991. During the 1991–92 growing season, the number of certified organic growers contracting with WFF increased to eighteen, and the resulting consignment of organic peas to Japan formed what one employee described as 'the most lucrative consignment of vegetables ever exported by Watties' (Campbell, 1996).

There were a number of reasons given by WFF for experimenting with organic production. The company had been using the perceived 'clean green' image of New Zealand products in marketing goods in Japan and two large consumer cooperatives in Japan began to enquire as to whether WFF could source totally organic products for them. In responding to this request, WFF reasoned that organic products could serve a number of useful purposes. First, they would be safe from non-tariff trade barriers dealing with chemical residues in food – something they were concerned about in the post-GATT trading environment. Second, organic products could be differentiated in the marketplace without requiring major production and processing changes for the company. Third, company strategists determined that the continued downward trend in bulk commodity prices necessitated a move toward higher value niche products. Finally, the company received enquiries from some of its established conventional growers who wished to convert to organics and were interested in whether WFF could cater for such products.

The combination of all these factors led to the decision by WFF to experiment with organic production and the success of the first consignment quickly pushed the organics programme to the forefront of the company's strategic planning. This support for the organics programme in WFF continued, even when Goodman Fielder divested itself of Watties and sold the Watties divisions to international food giant H.J. Heinz and Co. As part of the newly formed Heinz-Wattie food group in Australasia, WFF was encouraged to continue developing their organic interests.[4] The development was nurtured by a programme of support, extension and research titled 'Grow Organics With Watties' (GOWW). The company facilitating this programme with a full-time staff position, agricultural consultants, a newsletter for organic and interested growers and a specifically designated research budget to conduct research into organic production and management techniques.

The 1992/93 growing season saw a number of important developments. First, the company began to expand its organic activities to include the Gisborne, Manawatu and Hawkes Bay regions. This expansion was limited by an emerging difficulty in sourcing organic products as the first successful years of the programme had absorbed most of the extant organic producers. To encourage further organic production, WFF began to urge conventional growers to convert at least part of their farms to organic production. The company also began to concentrate on larger conventional growers. Alongside this recruitment strategy, WFF also extended the range of organic vegetables to include sweet corn, carrots and green beans. The result was that by the 1995/96 season (the fifth season of growing organic products), WFF had recruited fifty growers nationally with approximately 700 hectares of land in organic production for a predicted organic harvest of 10,000 tonnes of product. This represents around 7 per cent of WFF's total production (Campbell, 1996).

The WFF organics programme provided the major impetus behind organic exporting in Canterbury. However, once the WFF programme was established a number of other companies also began to engage in organic production. WFF's main vegetable processing competitors closely observed organic developments and approached key growers but could not find any growers willing to switch their contracts away from WFF for the 1996/97 season. Among the other companies now involved in organic production in Canterbury, many were directly influenced by the success of WFF, whilst others, in particular the baby food manufacturer Only Organics, initiated their operation independently of developments at WFF (Campbell, 1996). Consequently, it would be a mistake to assume that organic developments in Canterbury were restricted to the activities of one prominent company. Rather, a food production network began to emerge, and WFF and the other companies became interlinked in the emergence and operation of this network.

An emerging food network

In simply describing the nature of world demand for organic products and the emergence of companies catering to that demand, one can be drawn towards the erroneous assumption that such a development had emerged due to simple market forces of supply and demand. In the Canterbury case, this would produce a much simplified understanding of the complex contestations and alliances that have led to the emergence of organic exporting.

Marsden and Arce (1995) and Marsden (1996) suggest that food networks are not naturally occurring systems but are constructed and contested by many potential participants in these networks. Not only are the economic structures of food production, processing and distribution contested between parties, but the subjective nature of the food product itself is also an important factor structuring food networks. Foods have potential subjective qualities that are constructed according to the place of production (eg., the appellation system in France). These qualities can also be determined by the method of production, a factor that is particularly significant in the case of organic production. Cook (1994) has argued that the development of exotic fruits in the UK market was primarily driven by the symbolic qualities imbued in these fruits by wealthier consumers and supermarkets wished to stock such products for their symbolic value in marketing and store differentiation. Alongside these subjective qualities, food processors and distributors also desire particular qualities (eg., the desirability of 'firmness' in tomatoes). Fine et al. (1996) would go as far as to suggest that the organic[5] properties of food are what make food systems distinct from all other commodity systems. These organic qualities are manifested in both the subjectively constructed consumption and the economic structures of production, processing and distribution.[6]

The experience of a developing organic export industry in Canterbury reflects the dynamics embodied in the kinds of network discussed by Marsden and Arce (1995). Over the last five growing seasons, a number of points of tension have emerged in this nascent food network and these tensions have help shaped some of the outcomes of this development. In particular, issues relating to organic certification, grower relations with business intermediaries and infrastructural development have been crucial in the emerging network.

Certification of organic products Part of Marsden and Arce's (1995) analysis is the idea of 'quality' and how notions of 'quality' are constructed in food networks. For organic food networks, the issue of quality centres on systems of organic certification. In Canterbury, the organic certification standards established by the national organic certification agency BIO-GRO NZ. The formation (and review) of these standards has been negotiated between different groups within the organization (consumers and producers), between the national organization and the international organic farming body IFOAM,[7] and between WFF and BIO-GRO NZ.[8] To make the issue of 'quality construction' even more complex, WFF had its own sets of quality standards to apply, with national agricultural regulations and foreign market access regulations also applying to both growers and WFF. BIO-GRO NZ also found itself in the position of having to reconcile the demands of established organic growers and new, previously conventional, organic growers who had entered organic production through the Grow Organic With Watties programme. Many conventional growers were/are deeply suspicious about the involvement of WFF, while a number of newer growers were unsettled by many of ideological principles that had helped shape the organic agriculture movement in New Zealand. The development of workable compromises between the sometimes conflicting desires of these many parties has taken five growing seasons to develop. Potential conflicts are still arising in those parts of the emerging network relating to the subjective aspects of both the term 'organic' and 'sustainable agriculture.'

Grower relations with intermediaries Rural sociology has long been interested in the changing relations between family farms and off-farm agribusiness intermediaries. Theories of farm subsumption (Davis, 1980) posited that the involvement of family farms with capitalist agribusiness would lead to a progressive relinquishing of control over farm activities to the encroaching power of off-farm capital. More recently, food systems literature has tried to deconstruct the power relations inherent in food systems to incorporate an understanding of the contested and uneven nature of power between parties in food systems (Marsden,

1996). The Canterbury case study certainly exhibits these kinds of variable power relation between parties in the food system. The most unexpected effect of the internationalization of organic production is that organic producers developed greater freedom over the potential destination of their products. Growers that were previously only supplying the domestic market for organic food often found themselves having to supply only one cooperative outlet and in some cases, the shared ideological commitment to organic production was invoked to justify late payment to growers and the maintenance of a near monopsonistic position by local retail outlets (Campbell, 1996). The arrival of organic exporting significantly increased both the number of potential buyers for organic products, and the value of their products.

The restructuring of relations between organic growers and intermediaries had other effects that might be more in line with the expectations of subsumption theorists. The development of quality systems, as outlined above, and the competing needs of multiple parties in constructing quality criteria (of which organic growers form only one party), have certainly led to the imposition of complex production systems onto farm producers. It is partly for this reason that agricultural consultants working for WFF targeted skilled and managerially successful conventional farmers to join the Grow Organic With Watties programme. They perceived that it was these skilled farmers who could successfully manage these multiple demands to produce particular qualities in products.

These two, seemingly conflicting, tendencies in the emerging power relations between growers and intermediaries form an important and contested dynamic in the emerging organic exporting network. A final area of tension and contestation in the emerging network occurred in the development of infrastructure.

Infrastructure There are a number of infrastructural tensions that have emerged in the development of this food network. The subjective notions of 'quality' and the contestation of power in food networks are characteristics that create more complexity in the operation of food networks than might be found in the operation of bulk commodity markets. Consequently, the successful development of a food network involves contestable negotiations at many levels of the network. This places added pressure on the development and control of infrastructure – in particular the development of informal and formal business linkages – and also the development of sufficient size and scale in organic production to enable infrastructure to develop.

The first of these pressures relates to the size of the network. In the Canterbury case, one of the major problems facing the emerging network is the small number of growers producing organic products and the amount of time and skill required to enable growers to learn organic techniques. Many organic techniques involve strongly localized knowledges (see Kloppenburg, 1991, 1992) and these are not easy to integrate with traditional extension procedures. Another difficulty arises in coordination between different growers and companies. Broadacre organic production relies on the use of crop and stock rotations, necessitating that only a few products in each rotation in Canterbury can be supplied to WFF. Until other companies join the emerging network and increase the number of potentially compatible products in grower rotations, the financial security of growers will be threatened. Outside Canterbury, WFF has developed a full rotation of products which all can be absorbed in their Gisborne processing plant. This situation is not yet possible in Canterbury.

One major infrastructural outcome of the emerging food network has been the development of a new organizational structure to coordinate organic exporting activities. In 1995, the major companies involved in organic exporting, the two cooperating certification bodies and Tradenz (formerly the New Zealand Trade Development Board), created the Organic Products Exporting Group to help coordinate and strategise the development of the industry. Again, there were delicate political manoeuvres and compromises before all the parties could agree to a structure and agenda for this new organization.

Seeking an explanation: corporate window-dressing, corporate greening or green consumerism?

The case study outlined above raises a number of issues about how we understand organic agriculture in the current world food system. The following discussion will pursue one important issue in greater depth. Namely, how can we explain the development of organic exporting in the light of wider trends in global capitalism, food systems and sustainable agriculture?

By applying the theoretical idea of the food network it is possible to unravel the many competing and contingent factors that have led to the development of organic exporting in New Zealand. However, this case study is spatially restricted due to the scope of the original research project. The above sections are primarily focused on the development of an organic food network within New Zealand, and have not explored the relationship between this food network and more global trends of the kind posited by Friedland (1994), Le Heron and Roche (1995, 1996), Marsden and Arce (1995), and Watts (1996). In order to explore this relationship, a second body of literature must be engaged. This is the literature emerging from debates surrounding sustainable agriculture.

There are at least three distinct scenarios relating to the wider implications of agricultural sustainability and corporate capitalist activity, and these relate directly to the kinds of issues raised by the New Zealand case study. These three scenarios – corporate window-dressing, corporate greening and green consumerism – are an attempt, within the sustainable agriculture literature, to understand the emerging relationship between corporate capitalism and sustainable production systems. This section will review these scenarios and then concentrate on a fourth scenario – green protectionism – arguing that the New Zealand case study strongly supports this scenario in explaining why organic exporting has developed.[9]

The involvement of large agribusiness corporations in organic production in New Zealand is typical of a global trend towards the development of such relations. What is obvious in New Zealand is that the organic food industry has been profoundly internationalized. The predominantly domestic-oriented industry that existed prior to 1990 has become almost entirely export-oriented by 1996. The current situation in the United States could be better described as a process of 'de-localization' rather than internationalization as local suppliers become integrated with distant markets inside the US. This trend in organic food production has become significant enough to engage some of the leading commentators debating issues of sustainable agriculture. Friedmann (1993) is a pivotal figure in this emerging debate as she is the most prominent participant in both the discussion of sustainable agriculture and recent world food systems theory. Friedmann (1993) argues that the de-localization/internationalization of health foods and organic foods (generically – green food products), and the involvement of large agribusiness corporations in this process, is best explained by the scenario of corporate window-dressing.

Corporate window-dressing The wider literature on sustainable agriculture has not paid great attention to the direct involvement of corporate companies in organic production (or, in fact, any other form of sustainable agriculture). Instead, corporate businesses, and transnational corporations in particular, tend to be discussed in terms of how they are antithetical to the principles of sustainable agriculture. Friedmann (1993) clearly establishes a more specific theoretical critique through the idea of 'corporate window-dressing.' She argues that:

> the privileged can create small markets for plants cultivated and animals fed with fewer chemicals. They can even create profitable opportunities for agro-food enterprises, large or small, to bring them across the world. But this private solution ultimately reinforces the distance and durability that are unsustainable features of the present situation (Friedmann, 1993, p. 228).

Her rendition of the role of transnational corporations (TNC) is that even though corporate agriculture has become involved in production techniques developed within the sustainable agriculture movement, the overall structures of the Second Food Regime remain intact. Unlike Le Heron (1993), Le Heron and Roche (1995), Friedland (1994) and Watts (1996), who all argue that aspects of a Third Food Regime are starting to emerge, Friedmann (1993) insists that such a regime is still to develop and that if this regime is to be sustainable then the current structures of the Second Food Regime – with transnational corporations playing a major part – must be completely restructured. These differences in opinion over the status of a Third Food Regime have clear parallels to the related debate over the current or future status of post-Fordism as a global production regime.

Friedmann (1993) concludes that the incorporation of sustainable agriculture as a minor aspect of TNC activity simply serves to disguise the ongoing nature of structurally unsustainable production. These attempts to 'window-dress' corporate activities are intended to divert political and consumer attention away from the majority of TNC activities.

The corporate window-dressing scenario has clear links to a second line of argument best exemplified in Buttel (1992, 1993) (see also Bird, 1988; O'Connor, 1993; Allen and Sachs, 1993). Buttel (1992) argues that the actual terminology of sustainability has been appropriated by both agro-food capital and the capitalist state as a means of legitimizing or diverting attention away from existing production structures. James (1993) and Cocklin (1995) take this further by arguing that there is an important point of contestation within the development of sustainable agriculture and this can be identified in the discursive construction of differing notions of sustainability. James (1993) and Cocklin (1995) differ significantly from Buttel (1993) by taking a more contested view of the construction of discourses of sustainability. Buttel (1993) maintains, however, that corporate agribusiness has a controlling stake in the construction of this discourse. These two dynamics – TNC niche production sectors and corporate control of discursive definitions of sustainability – both fall clearly into what Friedmann (1993) considers to be 'corporate window-dressing.'

In conclusion, the corporate window-dressing scenario can be described as the 'radical' position on corporate involvement with 'sustainability.' Corporate window-dressing implies that the logic of capitalist activity eliminates the possibility of corporate involvement with sustainability having potentially beneficial outcomes for the system as a whole no matter what the surface appearance might be.

Corporate greening Having identified the radical position of 'corporate window-dressing,' a more moderate scenario can now be appraised. The scenario of corporate greening sees the potential involvement of agribusiness with sustainable agriculture in a slightly more positive light. Corporate greening is described by MacRae et al. (1993) as any attempt by corporations to meet demands for 'green' products – specifically to meet the demands of high-value niche consumers. In a similar vein to Friedmann (1993), McRae et al. (1993) argue that most attempts at corporate greening represent a partial rather than full restructuring of companies towards activities more acceptable to the green consumer. The clearest example of a large company that has 'greened' to a large extent is *The Body Shop* chain of stores owned by Anita Roddick. Most corporates, however, do not reach such degrees of greening and remain only partially committed to providing green products. Like Friedmann, MacRae et al. (1989) express some major concerns over the total sustainability of such activities, but the emphasis on companies reacting to consumer demands rather than developing strategies to deceive consumers provides an important distinction between the two positions. In the corporate window-dressing scenario, there is no incentive for companies to continue with conversion to greener activities once the deception is in place and the majority of their activities are successfully concealed. In corporate greening, however, the process is open-ended.

The pivotal distinction between these two positions is, therefore, the power of consumers to cause changes in corporate behaviour. It is around this crucial point in both food systems theory and wider theories of capitalist activity (Fine, 1994) that the relative merits of corporate window-dressing and corporate greening can be assessed.

Green consumerism Green consumerism refers to the rising levels of consumption of what Friedland calls 'fresh,' 'healthy,' food often produced from what are perceived to be 'sustainable' production systems. While many food systems theorists are prepared to accept that green consumers are influencing the structure of parts of the world food system, few are prepared to give them much weight. Buttel (1992) argues that the power of consumers will never be enough to convince many companies to convert fully to more sustainable activities. A similar position is put by both Fine et al. (1996) and Watts (1996) who consider that too much causal significance has been given by the environmental movement to the power of green consumers. Despite this, Watts (1996) and Friedland (1994) argue that much of the impetus towards an emerging regime of fresh and unprocessed foods has been driven by changing consumer demands in the First World. Marsden and Arce (1995) adopt a less negative position in their use of organic foods as a specific example of the way in which new consumer-driven niches/networks are emerging in the world food system. A similar tone of argument is made by Whatmore (1994) in her review of the restructuring of rural Europe. She argues that the agents of capital have been given too much weight compared to the power of emerging political movements and alliances. Again, the relative merits of contemporary or future green consumerism have strong similarities to the Fordist/post-Fordist debate.

These three scenarios represent a simplified attempt to review the current theoretical discussion of the place of corporate capitalism in sustainable agriculture. Each of these scenarios has potential relevance to understanding organic exporting in New Zealand, but, as the following section will argue, none of them is entirely convincing in explaining why this development has taken place. Instead, a fourth scenario – green protectionism – will be introduced and applied to the New Zealand case study.

Sustainable agriculture and green protectionism

The study of organic food networks emerging in New Zealand creates some difficulties for the supporters of the three above outlined scenarios of corporate involvement with sustainable agriculture. First, the discussion of corporate greening and corporate window-dressing may have mistakenly equated the move towards green products with only consumer pressures. Second, the pressures on corporates operating within an exporting country like New Zealand may be significantly different to the pressures experienced in countries where the main target market is domestic. This issue can be clearly identified in the post-GATT scenario of green protectionism.

The idea of green protectionism has arisen from debates over the nature of the world trading system in the post-GATT trading environment (McMichael and Myhre, 1991; McMichael, 1993; Le Heron, 1993). This scenario posits that countries will resort, for a number of reasons, to non-tariff barriers as they are forced to reduce regimes of tariff protection for their domestic agricultural sectors. Specifically, these non-tariff barriers might involve an increasing sensitivity to chemical residues in food or other environmental criteria.

The scenario of green protectionism has only, however, hovered at the margins of the sustainable agriculture debate.[10] The reason for this may be the current domestic-oriented structure of many corporates involved in producing green food products. To clarify this point further, the difference between 'delocalization' in Friedmann (1993) and internationalization in the New Zealand case is crucial. In Friedmann (1993), MacRae et al. (1993) and Buttel (1992), the companies being discussed in the context of corporate window-dressing or corporate greening are usually providing products for large-scale consumer markets inside their own national borders or within free trade areas (Canada/NAFTA, the United States, Britain/EU). If they are exporting green products then this is usually incidental to their core activities in their home markets and it is the dynamics within their home markets that tend to dictate corporate strategy. Consequently, the restructuring of trade barriers in the post-GATT global environment will have less effect on such companies than those companies operating in export-dependent countries like Australia and New Zealand.

The situation in New Zealand is undoubtedly driven by the perception that high-value niche markets in organic food do exist and may continue to grow. WFF have maintained that they have never yet been able to supply the full demand of only few purchasers in Japan, let alone try to cultivate other aspects of the Japanese or global market. However, the WFF experience in Japan reveals that two clear issues were influencing strategic planning in the company. The first was the high level of consumer demand, but the second was a growing concern over green protectionism in First World markets. The fact that the organics experiment was successfully applied in Japan first of all, clearly supports both these trends as Japan has a high level of consumer concern over chemical residues in food and some of the strictest non-tariff barriers against chemical residues in the world. As one Heinz-Watties director maintained:

> It might be argued that green marketing is possibly a consumer fad which might die away. Certainly there is evidence that sales peaked in 1988, with consumers abandoning premium priced green goods as the recession took hold. Nevertheless, a different phenomenon has been emerging (Garrett, 1995, p. 5).

This phenomenon is 'green protectionism' and WFF have successfully used organic produce as a 'keyhole' product to gain entry for both organic and conventional products into Japan.

The exact extent of green protectionism, both now and in the future, is impossible to gauge. It is clear, however, that some company strategists in export-dependent countries like New Zealand and Australia consider that the global conditions are in place for green protectionism to develop as a serious threat to export activity (see Garrett, 1995). One Heinz-Wattie manager in New Zealand described the scenario in these terms:

> Out of all that has been said and written about trading in the post-GATT world there has been one over-riding message and that is that as trade barriers disappear we can expect technical and, in particular, environmental restrictions to rise in their place. A surprising number of New Zealander's interpret this as 'business as usual'

> – after all don't we live in a 'clean green, environmental oasis' at the bottom of the world? However, most informed commentators seem to be agreed that it will be insufficient for NZ to rely on an *unsubstantiated* clean, green image in any market in the future (Davis, 1995, p. 2., emphasis in original).

WFF, as an exporter to the Japanese market, is already familiar with issues of green protectionism. Much of WFF's concern over quality control in organic production has been motivated by the serious degree to which Japanese borders are protected against food products bearing chemical residues. While there have been no instances of export consignments from WFF failing border inspections, other companies experimenting with organic production – in particular the NZ Kiwifruit Marketing Board and Apple and Pear Marketing Board – have already experienced green protectionism operating against their conventional products (particularly in the USA and EU).

Conclusion

The preceding discussion has raised four scenarios relating to the development of corporate export programmes in organic agriculture; corporate window-dressing, corporate greening, green consumerism and green protectionism. Contributors to the sustainable agriculture literature have tended to adopt either the radical 'window-dressing' or moderate 'corporate greening' scenario, and both of these have been linked with either a radical or moderate position on the power of green consumerism in restructuring capitalist food systems.

Two conclusions need to be drawn. First, that green consumerism is not the only process that is influencing the emerging structure of an organic food network in New Zealand. Instead, the emerging tendency for green protectionism in First World countries is having a significant impact on the way in which export agribusiness corporations are undertaking their operations. It is the twin scenarios of green consumerism and green protectionism that were the major factors behind Heinz-Wattie engaging in the search for products that were free of chemical residues. This led Heinz-Wattie to the complex task of alliance building with both the sustainable agriculture movement and a number of other companies interested in organic food. To undertake such a potentially uncertain venture, Heinz-Wattie required a powerful strategic incentive and that incentive was provided by the lure of green consumerism coupled with the threat of green protectionism.

The second conclusion to be drawn from the New Zealand case contrasts with the way in which 'greening' or 'window-dressing' has been analytically framed in the sustainability literature. Both the 'greening' and 'window-dressing' scenarios take the corporate business actor as the main unit of analysis and try to gauge the extent to which this actor has or has not adopted 'green' practices and products. As a TNC, Heinz-Wattie is involved in a multitude of often quite independent food networks, and the extent of 'greening' in one network might only vaguely be related to the operation or restructuring of other food networks which the company participates in. The theoretical scenarios of corporate 'window-dressing' or 'greening' consequently create a distorted focus on the corporate actor as the unit of analysis. This Chapter concludes that the concept of the food network must remain at the centre of analysis rather than one of the network's corporate participants. The New Zealand case provides a concrete example of an emerging green food network. The extent to which one corporate participant in this process was converted to green practices provides only part of the story. Future analysis will be better served by exploring the concept of food networks as proposed by Marsden and Arce (1995). It is in this kind of contested and contingent scenario that a better understanding of the emerging dynamics in the world food system can develop.

Notes

1. The author wishes to acknowledge the Foundation for Research Science and Technology in New Zealand for funding the research programme 'The Restructuring of Organic Agriculture in New Zealand,' the findings of which form the basis of this Chapter.
2. Globalization refers to the posited tendency for global economic and cultural activity to surpass the level of the nation/state and take on a global quality. Transnational corporations are usually implicated in this process (Coca Cola, MacDonalds, Hollywood film studios).
3. Organic food refers to the production of food according to standards that are developed by grower organizations at a national level. These standards determine the nature of acceptable farm inputs and generally reject the use of agri-chemicals. The international body IFOAM has recently been attempting to create unified international standards for organic food production. The ongoing concern for soil health and environmental outcomes of farm production place organic production clearly within the wider agricultural sustainability movement in the First World.
4. The chairman, president and chief executive officer of H.J. Heinz, Tony O'Reilly, confirmed an interest in organic foods on a global scale with the purchase of two organic food processing companies – Italian company Fattoria Scaldasole SpA (annual sales of US$28 million) and US company Earth's Best (annual sales US$1 billion). Both these purchases took place in one week in early 1996.

5. Meaning physical/biological properties, not to be confused with organic food.
6. For a counter argument see Watts (1996).
7. Mainly due to some BIO-GRO NZ standards being more stringent than the IFOAM equivalents.
8. The development of workable procedures for inspection had to be negotiated between parties to enable WFF to streamline grower participation and production while BIO-GRO NZ inspectors retained their status as independent certifiers.
9. In the US and EU-based literature, organic agriculture is seen as merely one of a number of production systems that might be considered 'sustainable agriculture.' In New Zealand, the sustainable agriculture social movement has given more weight to organic agriculture than other production systems (eg., permaculture) in defining itself. While there have been many attempts by different parties (policy makers, agricultural scientists, fringe political parties, economists) to redefine 'sustainability,' (Cocklin, 1995), the sustainable agriculture movement has retained its own definitions of sustainability and carried such definition(s) into the emerging food network around organic agriculture. Consequently, in New Zealand, the debates and conflicts around organic agriculture form a useful picture of the wider discussion of agricultural sustainability. This tight focus on organic systems is not so applicable for the US and EU-based literature discussed in this section, and consequently, the term 'green' food is often used to incorporate a variety of products from what might be claimed to be 'sustainable' agriculture systems.
10. This is not to imply that commentators interested in sustainability have not speculated at length about the consequences of the completion of the Uruguay Round of GATT, rather, these ideas have not been pivotal in the specific discussion of sustainable agriculture and corporate activity.

References

Allen, P. and Sachs, C. (1993), 'Sustainable Agriculture in the United States: Engagements, Silences, and Possibilities for Transformation,' in Allen, P. (ed.), *Food for the Future: Conditions and Contradictions of Sustainability*, John Wiley and Sons: New York.

Arce, A. and Marsden, T. (1993), 'The Social Construction of International Food: A New Research Agenda,' *Economic Geography*, Vol. 69, pp. 293–311.

Bird, E. (1988), 'Why "Modern" Agriculture is Environmentally Unsustainable: Implications for the Politics of the Sustainable Agriculture Movement in the U.S.,' in Allen, P. and Van Dusen, D. (eds), *Global Perspectives on Agroecology and Sustainable Agricultural Systems*, Agroecology Program – University of California: Santa Cruz, CA.

Buttel, F.H. (1992), 'Environmentalisation: Origins, Processes, and Implications for Rural Social Change,' *Rural Sociology*, Vol. 57, No. 1, pp. 1–27.

Buttel, F.H. (1993), The Production of Agricultural Sustainability: Observations from the Sociology of Science and Technology, in Allen, P. (ed.), *Food for the Future: Conditions and Contradictions of Sustainability*, John Wiley and Sons: New York.

Buttel, F.H., Larson, O. and Gillespie, G. (1990), *The Sociology of Agriculture*, Greenwood Press: New York.

Campbell, H. (1994), *Regulation and Crisis in New Zealand Agriculture: The Case of Ashburton County, 1984–1992,* Unpub. Ph.D thesis in Sociology, Charles Sturt University, Australia.

Campbell, H. (1996), *Recent Developments in Organic Food Production in New Zealand: Part 1, Organic Food Exporting in Canterbury,* Studies in Rural Sustainability Research Report No. 1, Department of Anthropology, University of Otago.

Cocklin, C. (1995), 'Agriculture, Society and Environment: Discourses on Sustainability,' *International Journal of Sustainable Development and World Ecology,* Vol. 2, pp. 240–256.

Cook, I. (1994), 'New Fruits and Vanity: Symbolic Production in the Global Food Economy,' in Bonnano, A. et al. (eds), *From Columbus to ConAgra: The Globalization of Agriculture and Food*, University Press of Kansas: Lawrence.

Davis, J.E. (1980), 'Capitalist Agricultural Development and the Exploitation of the Propertied Laborer,' in Buttel, F.H. and Newby, H. (eds), *The Rural Sociology of the Advanced Societies: Critical Perspectives.* Allanheld: Osmun, Montclair.

Davis, S. (1995), 'Some Plant Protection Issues for Wattie Frozen Foods Ltd.' Opening Address to the 48th New Zealand Plant Protection Conference, Hastings, 8 August 1995.

Fine, B. (1994), 'From Political Economy to Consumption,' in Miller, D. (ed), *Acknowledging Consumption: A Review of New Studies*, Routledge: London.

Fine, B., Heasman, M. and Wright, J. (1996), *Consumption in the Age of Affluence: The World of Food,* Routledge: London.

Friedland, W.H. (1980), 'Technology in Agriculture: Labour and the Rate of Accumulation,' in Buttel, F.H. and Newby, H. (eds), *The Rural Sociology of the Advanced Societies: Critical Perspectives*, Allanheld: Osmun, Montclair.

Friedland, W.H. (1984), 'Commodity Systems Analysis: An Approach to the Sociology of Agriculture,' *Research in Rural Sociology and Development*, Vol. 1, pp. 221–235.

Friedland, W.H. (1994), 'The New Globalization: The Case of Fresh Produce,' in Bonnano, A. et al. (eds), *From Columbus to ConAgra: The Globalization of Agriculture and Food*, University Press of Kansas: Lawrence.

Friedland, W.H., Barton, A. and Thomas, R. (1981), *Manufacturing Green Gold: Capital, Labour, and Technology in the Lettuce Industry*, Cambridge University Press: Cambridge.

Friedmann, H. (1990), 'The Origins of Third World Food Dependence,' in Bernstein, H. et al. (eds), *The Food Question: Profits Versus People,* Monthly Review Press: London.

Friedmann, H. (1991), 'Changes in the International Division of Labor: Agri-Food Complexes and Export Agriculture,' in Friedland, W.H. et al. (eds), *Towards a New Political Economy of Agriculture*, Westview Press: Boulder.

Friedmann, H. (1993), 'The Political Economy of Food,' *New Left Review*, No. 197, pp. 29–57.

Friedmann, H. and McMichael, P. (1989), 'Agriculture and the State System,' *Sociologia Ruralis*, Vol. 24, No. 2, pp. 93–117.

Garrett, S. (1995), 'Opportunities and Challenges for Agriculture,' Paper presented to Agricultural Economics Society Conference, NZ 1995.

Goodman, D. and Watts, M. (1994), 'Reconfiguring the Rural or Fording the Divide? Capitalist Restructuring and the Global Agro-Food System,' *Journal of Peasant Studies,* Vol. 22, No. 1.

James, A. (1993), 'Eating Green(s): Discourses of Organic Food,' in Milton, K. (ed.), *Environmentalism: The View from Anthropology*, Routledge: London.

Jarosz, L. (1996), 'Working in the Global Food System: A Focus for International Comparative Analysis,' *Progress in Human Geography,* Vol. 20, No. 1, pp. 41–55.

Kenney, M., et al. (1989), 'Midwestern Agriculture in US Fordism,' *Sociologia Ruralis,* Vol. 29, No. 2, pp. 131–148.

Kenney, M. et al. (1991), 'Agriculture in U.S. Fordism: The Integration of the Productive Consumer,' in Friedland, W.H. et al. (eds), *Towards a New Political Economy of Agriculture,* Westview Press: Boulder.

Kloppenburg, J. (1991), 'Social Theory and the De/Reconstruction of Agricultural Science: Local Knowledge for an Alternative Agriculture,' *Rural Sociology*, Vol. 56, No. 4, pp. 519–548.

Kloppenburg, J. (1992), 'Science in Agriculture: a Reply to Molnar, Duffy, Cummins, and Van Santen and to Flora,' *Rural Sociology,* Vol. 57, No. 1, pp. 98–107.

Lawrence, G. (1996), 'Contemporary Agri-Food Restructuring: Australia and New Zealand,' in Burch, D., Rickson, R. and Lawrence, G. (eds), *Globalisation and Agri-Food Restructuring: Perspectives From the Australasia Region,* Avebury: Aldershot.

Le Heron, R. (1989a), 'A Political Economy Perspective on the Expansion of New Zealand Livestock Farming, 1960–1984: Part One. Agricultural Policy,' *Journal of Rural Studies,* Vol. 5 No. 1, pp. 17–32.

Le Heron, R. (1989b), 'A Political Economy Perspective on the Expansion of New Zealand Livestock Farming, 1960–1984: Part Two. Aggregate Farmer Responses – Evidence and Policy Implications,' *Journal of Rural Studies,* Vol. 5, No. 1, pp. 33–43.

Le Heron, R. (1993), *Globalised Agriculture: Political Choice*, Pergamon Press: Oxford.

Le Heron, R. and Pawson, E. (1996), *Changing Places: New Zealand in the Nineties,* Longman Paul: Auckland.

Le Heron, R. and Roche, M. (1995), 'A "Fresh" Place in Food's Space,' *Area,* Vol. 27, No. 1, pp. 23–33.

Le Heron, R. and Roche, M. (1996), 'Eco-Commodity Systems: Historical Geographies of Context, Articulation and Embeddedness under Capitalism,' in Burch, D., Rickson, R. and Lawrence, G. (eds), *Globalisation and Agri-Food Restructuring: Perspectives from the Australasia Region,* Avebury: Aldershot.

Le Heron, R., Roche, M. and Anderson, G. (1989), 'Reglobalisation of New Zealand's Food and Fibre System: Organisational Dimensions,' *Journal of Rural Studies,* Vol. 5, No. 4, pp. 395–404.

Le Heron, R., Roche, M. and Johnston, T. (1992), *Pluriactivity in New Zealand Agriculture in the 1980's: A Benchmark Study of Livestock and Fruit Agro-Commodity Systems,* Research Paper, Department of Geography, Massey University.

Marsden, T. (1996), 'Rural Geography Trend Report: The Social and Political Bases of Rural restructuring,' *Progress in Human Geography*, Vol. 20, No. 2, pp. 246–258.

Marsden, T. and Arce, A. (1993), *Constructing Quality: Globalisation, the State and Food Circuits,* Globalization of Agriculture and Food Working Paper Series, University of Hull, The Agricultural University: Wageningen.

Marsden, T. and Arce, A (1995), 'Constructing Quality: Emerging Food Networks in the Rural Transition,' *Environment and Planning A*, Vol. 27, pp. 1261–1279.

McMichael, P. (1993), 'World Food System Restructuring Under a GATT Regime,' *Political Geography*, Vol. 12, No. 3, pp. 198–214.

McMichael, P. and Myhre, D. (1991), 'Global Regulation v's. the Nation-State: Agro-Food Systems and the New Politics of Capital,' *Capital and Class*, Vol. 43, pp. 83–105.

MacRae, R.J., Henning, J. and Hill, S.B. (1993), 'Strategies to Overcome Barriers to the Development of Sustainable Agriculture in Canada: The Role of Agri-Business,' *Journal of Agricultural and Environmental Ethics*, Vol. 6, No. 1, pp. 21–51.

McErlich, A. (1994), 'The Development of Market Opportunities for Organic Frozen Vegetables to Japan,' *Proceedings of the Nutrition Society of New Zealand*, Vol. 19, pp. 32–36.

O'Connor, J. (1993), 'Is Sustainable Capitalism Possible?' in Allen, P. (ed.), *Food for the Future: Conditions and Contradictions of Sustainability*, John Wiley and Sons: New York.

Watts, M. (1996), 'Development III: The Global Agrifood System and Late Twentieth-Century Development (or Kautsky *redux*),' *Progress in Human Geography*, Vol. 20, No. 2, pp. 230–245.

Whatmore, S. (1994), 'Global Agro-Food Complexes and the Refashioning of Rural Europe,' in Amin, A. and Thrift, N. (eds), *Globalisation, Institutions and Regional Development in Europe*, Oxford University Press: Oxford.

4 The struggle surrounding the distribution of food in Norway

Berit Nygård and Oddveig Storstad

Introduction

The question of who should take care of the distribution of food is one of the hottest conflict areas in the Norwegian agro-food sector. Both the producers and the retailers want to control distribution. The struggle over sales concerns the producers' wish to continue their own distribution, and the wholesalers' and chainstores' demands to establish coordinated distribution from their own central storage. We pose the question of what possibilities the various actors have for enforcing their demands.

Distribution is the most important link between the producer and the consumer. Restructuring the distribution link is part of a broader restructuring process in the agro-food sector. Our study joins the tradition of rural sociological research that is centered around the food chain.

Throughout the advanced industrial world fundamental changes are taking place in the production and distribution of food. A number of studies have dealt with the restructuring of the agro-food sector (Ward, 1990; Friedland, 1991; Goodman and Redclift, 1991; Munton, 1992; Wallace, 1992; Wrigley, 1993; Flynn, Marsden and Ward, 1994; Almås and Ward, 1994; Blekesaune et al., 1995). In a broad definition of the agro-food system, these studies encompass the agricultural sector, those sectors supplying agriculture with inputs, food processing, manufacturing, distribution and food retailing (Friedman, 1982). Almås and Ward (1994, p. 2) stress that it is increasingly inappropriate to assess change in individual sectors (agriculture, food processing, food retailing) without reference to the ways that economic sectors are organized and linked together. Our study focuses, in particular, on the interests that are tied to the distribution of food.

Background

The struggle surrounding the distribution of food in Norway must be contextualized as part of the development of Norwegian food retailing during recent years. The number of grocery stores in Norway (as in a number of countries) has been drastically reduced during the past decade, from approximately 8,000 in 1982, to well under 6,000 stores in 1993. At the same time, major portions of the grocery trade have been concentrated into chains. Sixty per cent of the grocery trade in Norway, in August 1994, took place through the three chainstores: the REMA group, the Hakon group and the Consumer Cooperative (NKL) (Blekesaune et al., 1995). The remaining stores, which are not part of any of these chains, have organized an alternative cooperative for buying, which is called Norwegian Retailers. These independent stores accounted for 37 per cent of the total turnover in August 1994 (Blekesaune et al., 1995). A certain degree of vertical integration has taken place within individual chainstores; in other words, the connection between the retailers and the wholesalers has become closer, either through ownership or through special conditions for cooperation. This strong development of the chains has been a necessary condition for the transition from direct distribution to wholesale distribution (Blekesaune et al., 1995). By tieing the wholesalers more closely to their own companies, the retailers have achieved greater control over the supply of products. This increased power of the chains has strengthened the retail link in the supplier/producer relationship. Another central trend is that the grocery trade has become more international and the competition has become stiffer. All together, this may provide a background for conflicts and struggles for power on different levels.

A single product goes through many different links on its way from the producer to the consumer. Put simply, distribution comprises the transfer of wares and services from where they are produced to the locations where they are consumed (Jacobsen and Dulsrud, 1994, p. 152). Among the various types of distribution we may generally distinguish between direct and indirect forms. Direct or separate distribution means that the producer undertakes most of the tasks himself and is in direct contact with the retailers. With indirect distribution, the producer uses an intermediary to market his products, delivering his wares to the middleman's terminal, from which the middleman delivers the wares to the stores. Indirect distribution can take place in different ways. Jacobsen and Dulsrud (1994, p. 157) distinguish between coordinated distribution and wholesale distribution. Coordinated distribution – the most prevalent – includes cooperation in distribution for both independent businesses and producers (as with SAMDI – the coordinated distribution in the Farmers' Cooperative), together with wholesalers and various forms of cooperation between all of these. The term 'wholesale distribution' includes forms of distribution which are taken care of by the wholesaler or the retailer.[1] There is little direct sale of farm products in Norway. Most of the products are sold through various farmers' cooperatives which together comprise the Farmers' Cooperative. The Farmers' Cooperative in Norway is divided into sales, purchasing and breeders' cooperatives and consists of sixteen organizations that cooperate through the Farmers' Cooperative General Office. The sales organizations of the cooperatives were originally established to implement the receiving, processing, distributing and marketing of farmers' products. The Farmers' Cooperative still has a very strong position in Norway.

The pattern of distribution in Norway is different from other European countries in that the proportion of wares distributed directly to the grocer is rather large (Jacobsen and Dulsrud, 1994, p. 158). Sixty per cent of the deliveries to the retailer are covered by direct distribution, and 40 per cent through coordinated distribution by means of the grocery chains' own central warehouses. The geographical conditions in Norway would lead one to expect a much higher percentage of coordinated distribution. This is clear from comparison with the UK, for example, where coordinated distribution is close to 100 per cent.

The greatest proportion of separate or direct distribution of products is found in the group that includes bakery products, dairy products, meat, fruit and vegetables. There is a long tradition of the supplier distributing his own products, for both historical and technical reasons. The result of this has been that the wholesaler often lacks both the competence and the capacity to organize the distribution of fresh products, which require certain procedures regarding handling, transport, cooling, storage, hygiene, etc. (Jacobsen and Dulsrud, 1994, p. 158).

The strong concentration on chainstores, and the close ties between the wholesaler and the retailer, have resulted in 'closed' channels for products, which makes it difficult for new actors to enter the arena. This has also led to a power shift in the grocery trade from producers and wholesalers to the retailers, creating an entirely new situation for producers (Blekesaune et al., 1995). Whereas producers formerly negotiated with many retailers, they now only have to negotiate with the major chains. Thus it is even more important to secure shelf space in their stores. The major grocery chains control such a large portion of the Norwegian grocery trade that losing a contract with a chain would be a catastrophe for a producer. In order to ensure continued markets for Norwegian farmers' products, the Farmers' Cooperative has planned to establish a nationwide coordinated distribution of farm products. These plans challenge the Farmers' Cooperatives own customers (the grocery chains) since, in today's situation, the chains have control over major sections of distribution. We have studied how these plans are being received by the other actors in the food market. As we shall see, there is a clear conflict between the retailers and the producers' interests, and we will look more closely at the underlying factors in the struggle surrounding the distribution of food in Norway.

Theoretical framework and method

We decided to use concepts from games theory as a framework for our interpretation in order to achieve a better understanding of the mechanisms which seem to create actual relationships between the players in the food market. From this starting point, we can interpret the players alternatives for action, the consequences of each alternative for action and the players' preferences and information within their models of rationality (Hovi and Rasch, 1993). As suggested by its name, this is a 'game' in which the players also consider the other players' choices of action. The final results of their own actions are dependent on the alternatives for action that the other players choose. In other words, it is necessary to be able to 'interpret the game.' We have here a game where each participant operates as a player who uses all possible (rational) advantages but, at the same time, there are limitations in our ability to account for all of the situations that may occur in the game. This is because we are talking about a dynamic game, where the framework (the rules of the game) within which the game is being played is constantly changing. One of the characteristics of the game is that the players do not necessarily make their moves simultaneously, making it possible to see both ahead and back in time when making decisions about the choices of action.

As previously mentioned, the Farmers' Cooperative has plans to establish common distribution of fresh produce. The plans are mainly in response to the stores' wish to have as few deliveries by as few trucks as possible in the course of a day. As a player in this 'game,' the Farmers' Cooperative is making its move in accordance with the moves (choice of action) made by the other players earlier in the game.

As a result of the process of vertical integration which has been taking place within the grocery sector, the retailers themselves have made moves to take control of distribution (Blekesaune et al., 1995). They also want to integrate the farm products in their own distribution networks, and thus do not welcome the plans of the Farmers' Cooperative. Importation of farm products is limited in

Norway, but the national farm policies are under pressure and undergoing change. In the near future, obligations under the WTO agreement (formerly GATT) may require Norway to open its market to imports of foreign farm products. This means that in this 'game' there is also the possibility of a threat from one of the players toward another (the chains can, for example, threaten to sell products from foreign producers, if this would be advantageous economically). It is also possible to punish or reward each other for acting in a specific manner. 'If you (the farmers) don't play according to our (the retailers) rules, we won't play with you any more – we can find someone else who will play by our rules.' In a situation like this it is important to ask: Are the pressures applied by one of the participants in the 'game' credible, or only empty threats?

One of our questions concerns the credibility of these threats, and what alternatives retailers have in choosing sources of fresh produce other than the Farmers' Cooperative. In this context it is important to look at the balance of power between the various participants in the 'game.' The balance of power within the food system has changed over time, and often been unstable. Over the past few years, the growth of the grocery chains has resulted in an increase in the power of the retailers in the food system, while the primary producers and the processing industry have lost much of their influence (Blekesaune et al., 1995).

In a market where different actors perform, there will also be differences in the degree of power attached to the various roles. Galbraith's (1967) alternative to economic models based on the concept that supply and demand will regulate themselves through free competition, emphasized how the market is regulated by powerful actors being met by countervailing power. Power in the market may be exercised by strong buyers against weak sellers (for example in the employment market), as well as strong sellers against weak buyers (for example in the consumers' market). Retailers first came into the picture as a link between sellers and buyers in the consumers' market. The retailers presented themselves as acting on behalf of consumers in establishing opposition to the powerful producers. Galbraith (1967) regarded the growth of grocery chains as a concrete result of the retailers' strategies to achieve a balance by challenging the strong economic power of the market. Nearly thirty years after Galbraith launched his theory, the situation is different. Now the retailers are the ones with the greatest power. In this context, the Farmers' Cooperative's plans for coordinating distribution may be regarded as a countervailing power.

In order to shed light on the various interests that are tied to the distribution of food, we chose to use two sources of information: qualitative informant interviews and studies of newspapers and trade periodicals. Qualitative informant interviews can take various forms, as described for example in Miles and Huberman (1984).

Informant interviews were carried out during the autumn of 1994 in conjunction with a comparative study of structural changes in the food sector (Blekesaune et al. 1995). Key persons within the Norwegian grocery sector, as well as from the Danish and Swedish grocery sector and food industry, were interviewed. Here we concentrate mainly on interviews with representatives from the Norwegian grocery branch.

Initially we wanted to interview key informants from all of the three dominant chains in the Norwegian food retail sector (the Consumer Cooperative, the REMA group and the Hakon group), as well as representatives from Norwegian Retailers. In order to find these key informants, we first contacted the various businesses by telephone, and were given the name of an executive whom we could contact directly in order to request an interview. With one exception, we received a positive response to our requests about informant interviews in the Norwegian food retail sector.[2] Ten interviews were carried out in Norway, Sweden and Denmark. The four interviews with representatives from the Norwegian grocery sector form the basis for this Chapter.

The interviews took place in person and, since more than one interviewer was involved, a schedule was useful for ensuring that the interviews followed the same pattern. We chose to set up an interview guide with specific, predetermined questions to be certain that we would acquire information about the topics in which we were interested. At the same time, this allowed a more flexible approach which could be adapted according to need. The interview guides varied somewhat, depending on whom we were interviewing, and the type of business the informant represented. The questions about product distribution as such were part of a more extensive list of topics basically concerned with today's situation in the food retail sector, along with considerations and strategies for the future.

Most of the interviews were, with the consent of the respondents, recorded on tape and later transcribed in their entirety. In the analysis, we will use quotations from the interviews to illustrate the grocery chains' attitudes to the distribution of food products.

In addition to the informant interviews, newspapers and trade periodicals were also important sources of data. We found it very useful to study previous years' issues of *Dagligvarehandelen* (The Grocery Trade) and *Fritt Kjøpmannskap* (Free Merchandising).[3] While these two trade papers gave us useful background material for developing the interview guide, they also helped keep us oriented with respect to developments within the branch after the interviews had taken place. The periodical *The Farmers' Cooperative*, has been a key source of information about the Farmers' Cooperative's interests and their plans to establish nationwide coordinated distribution of fresh produce.

In addition, we used daily newspapers as a source of data. We systematically collected articles from various newspapers (mainly *Dagens Næringsliv* – Daily Business, *Nationen* – a national newspaper mainly dealing with farmer issues, and *Bondebladet* – The Farmers' News). We refer to these sources where we have used them. We should point out that although we used newspapers and trade periodicals as sources of data for the project, they are not intended as primary research sources. They are, rather secondary sources, in contrast to the interview data which are primary sources. Secondary sources have the weakness of being 'copies' – we must relate to information that is communicated through another person (source). This information can therefore be 'coloured' by that person's/newspaper's interpretations; therefore it is important that we reveal our source.

Qualitative informant interviews also have their limitations. One of the limits depends on whether one accepts 'texts' as an accurate reflection of individual ideas. At the same time, the primary data (interviews) have the disadvantage that they were collected early (autumn 1994), at a time when the Farmers Cooperative's plans for coordinated distribution of fresh produce were comparatively new. In general, one might say that primary sources are the strongest, while secondary sources are poorer (Rosengren and Arvidson, 1983). Evaluating the use and combination of primary and secondary sources, we would maintain that the two complement each other.

Results

The shift in the organization of food distribution the past two years suggests that fundamental changes are taking place in the sharing of distribution tasks (Jacobsen and Dulsrud (1994, p. 161). Changes in the structure of sales channels often lead to vertical conflicts in the distribution chain (Blom et al., 1985), when new actors try to gain entry into established channels, or when several actors in a sales channel try to eliminate others. The various actors have different interests in distribution. The suppliers claim that given the way wholesale distribution is organized today, direct delivery is the cheapest or most cost-effective form of distribution. If wholesale distribution increases the cost of the products, prices will increase and this in turn will reduce demand and total turnover. The arguments against giving up direct distribution have been strongest from the suppliers of fresh wares, such as meat and dairy products. According to the suppliers, there are several factors which make direct distribution superior to wholesale distribution (refrigerated trucks, avoiding reloading, etc.) (Jacobsen and Dulsrud, 1994). The chains and wholesalers, on the other hand, claim that direct delivery means extra costs for the consumer, and that this form of distribution is not rational or efficient for the individual store's receiving routines. We shall now look more closely at the grocery chains' interests in the distribution of food.

The grocery chain's interests in food distribution

As few daily deliveries as possible The retailers prefer having as few deliveries as possible during the course of a single day. A study published by the *Dagligvarehandelen* (The Grocery Trade magazine) in October 1993 shows that the grocery stores would like to reduce the number of deliveries per week, in keeping with the wishes of the leaders of the chains. More wares through the wholesale link will increase efficiency and save on costs. The reduction of costs which can be attained through vertical integration gives the most well-integrated retail chains a competitive lead over the less well integrated.

Coordinated distribution through own wholesale network In the future, the retailers would like to get farm products delivered to their own central warehouses, to be transported from there in their own clearly marked trucks. Again, the motive is to have power and control over the products.

The development toward integrated and closed channels puts many suppliers in a dilemma. They can choose to give up their own distribution to gain access to the market; this would reduce their independence and the possibility of having marketing power through their own distribution. On the other hand, the suppliers can choose to maintain their own distribution, to have a certain advantage in the competition. Then, however, they risk losing access to the market (Jacobsen and Dulsrud, 1994, p. 170–1).

Gartnerhallen is the business/enterprise within farm cooperatives which first experienced this vertical integration process. In 1993, REMA 1000 discontinued its contract with Gartnerhallen as its supplier of fruit and vegetables, and entered a contract with Norgesfrukt A/S (Norwegian Fruit, Inc.) as their sole supplier. The termination of the contract was not primarily the result of dissatisfaction with Gartnerhallen, but the wish to have a closed system of distribution. This illustrates the great power the retail chains have attained over the producer and distribution links during the past few years.

Reduction in costs By making distribution more efficient, the grocery chains can reduce their costs, something which is regarded as the most important challenge in today's situation. The food and beverage industries are obvious targets for making the distribution from the producer to the retailer more cost effective (*Fritt Kjøpmannskap* 12-09-94).

In order to achieve the most rational distribution possible, the grocery chains are preoccupied with increasing their so-called wholesale share, in other words with moving from direct distribution to coordinated distribution from the chains' own central warehouses. That is in part what has happened in connection with the chocolate producer Freia and some of the suppliers who deliver their products directly to the stores.

> We want them to deliver directly to our storage, and we will take care of the distribution ourselves, because it is cheaper and better, on the whole. It is also better for society. It is cheaper to drive one vehicle with all the products to the stores than having each supplier deliver with his own truck. (Interview Hakon group)

Rationalizing distribution is emphasized more when we consider further expansion of the chains. There is not much room for further expansion in the Norwegian market. The Hakon group, which currently has a market share of 24 per cent, refers to further expansion of up to a third, but that is not regarded as the most important challenge for the future.

> I believe we can increase some more, but not much more. I believe the challenge for us lies more in taking care of the network of stores we already have, doing things better and doing things more efficiently and reducing costs. This is our challenge in the time to come. (Interview Hakon group)

The reduction of costs is, then, seen as the most important challenge of the future. As a move towards cutting distribution costs, the Hakon group has entered into cooperation with ICA from Sweden. Hakon has introduced its low-price RIMI concept to Sweden, and ICA has established its supermarket concept in Norway. These two companies have furthermore established a purchasing cooperative, Viking Retail Alliance, together with Kesko of Finland. One of the most important reasons for starting this cooperative, was that by increasing the volume they would be stronger in their negotiations with foreign suppliers (Blekesaune et al., 1995). In other words, this is integration on a Nordic basis where their negotiating position has become stronger and – in the future – with integration of distribution as a goal. The imminent decrease in the protection from imports already indicated by the GATT agreement presents

Norwegian food producers with new challenges. A purchasing cooperative (such as the Viking Retail Alliance) will, because of its volume, be able to achieve such favourable agreements with foreign producers that, in spite of customs duties, it will outclass the competition from products produced within the country.

The Consumer Cooperative (NKL) also points to improvement of distribution as an important step in improving its ability to compete.

> It is unbelievable when you think that 15 trucks drive up through Gudbrandsdalen on Monday morning in order to fill up a small shop up in the valley. We have become used to this type of arrangement – but the competitive situation is pressuring us to rid ourselves of it. (Interview NKL)

This strong focus on cost reduction and making distribution more efficient has also made the grocery chains look at their own suppliers in a different light. The most important thing used to be to squeeze every penny out of the supplier in order to get a lower price. Price is still a factor, but in addition there is also a wish to achieve a more rational flow of goods through cooperation with the supplier. Retailers are entering partnership agreements with many of the large suppliers, and are further reducing their costs in this way by making distribution more efficient. In other words, retailers are demanding that their suppliers join them in projects designed to create cost advantages along the flow of products.

The grocery chains' wish to control as much of the distribution as possible in order to reduce costs. At the same time, this also opens doors for more flexible solutions that enable them to choose suppliers outside the farm cooperative. Although many chains state that they prefer to use Norwegian suppliers, it is also clear that they will not do this at any price.

> We place demands on competition because we see that other actors in the trade are buying in products where they are cheaper, so then we cannot let ourselves be outclassed by the competition because we hold on to an idealistic concept of what we are doing. (Interview NKL).

This is a good illustration, seen in the light of games theory, of the way a player acts according to his judgment of how competitors will act, even though this may be in conflict with his own ideals.

The Farmer Cooperative's plans for coordinated distribution of fresh produce

The challenges facing the primary producers today, with the dominant chains and increased internationalization, demand a new way of thinking in order to achieve cooperation. The farm cooperative has several strategies from which to choose in its confrontation with the powerful grocery chains. The Godt Norsk (Good Norwegian) campaign is one strategy, coordinated distribution is another. The idea of coordinated distribution is not new to the ranks of the Farmer Cooperative.[4] However, the idea of nation-wide coordinated distribution developed only during the last few years. In 1994, Norske Meierier (Norwegian Dairies), Norsk Kjøtt (Norwegian Meats), Gartnerhallen (Fruits and Vegetables) and Norske Eggsentraler (Norwegian Eggs) established the company Landbrukets Ferskvaredistribusjon A/S (The Farmers' Fresh Produce Distribution, Inc.). The purpose of this company is to extend the farmers' cooperatives to the level of distribution.

Reduce costs and ensure market access The farm cooperative's plans for coordinated distribution of fresh produce have been motivated by several factors. First, cooperation in this area has a great potential for reducing costs – by an estimated NOK 450 million per year (Hertzenberg, 1994). This saving is important at a time when the transfer of money from the federal budget to the farm industry is increasingly being questioned. The other, and perhaps most important reason, is the strategic element contained in a common distribution of goods, which may be decisive for securing continued access to the produce market for Norwegian farmers. The common distribution of fresh produce will be put into practice by establishing 16–22 common freight terminals spread out among various regions, from which produce will be transported in a single truck to stores in the surrounding region. This is expected to result in approximately 300 fewer trucks.

In Galbraith's (1967) terms, the Farmer Cooperative's plans for coordinated distribution of farm produce can be seen as an attempt to build up countervailing power. By coordinating the delivery of milk, meat, eggs, fruit and vegetables, the Farmer Cooperative secures access to the market for its products. The cooperative organizations were originally established for just this purpose: a means of receiving, processing, distributing and marketing farmers' products.

How are the farm cooperative's plans for coordinated distribution of fresh produce met by the other actors in the food chain?

While both the grocery chains and the Farmer Cooperative want to move toward a greater degree of coordinated distribution, these two actors define this term in different ways. The fight in the sales channel is between the suppliers' wish to maintain their own distribution and the wholesalers' and chains' demands to achieve solutions through coordinated distribution.

What possibilities do these different actors have to enforce their demands? As previously mentioned, retailers have the most power. Does this mean that they will set the rules of the game? The challenge for the farm cooperative lies in being so efficient in their distribution that the other actors will not find it economically interesting to take it over. Coordinated distribution is a prerequisite for achieving such a position. The possibilities for carrying out these plans will depend on how they are met by the other actors in the food chain.

The Farmer Cooperative's plan to coordinated distribution of fresh produce by concentrating all their deliveries into a single one in response to the stores' wish to have as few daily deliveries as possible. The Farmers' Cooperative's coordinated distribution plans sound like a step backward for the retail chains, whose goal is to run the distribution from their own central warehouses. They want coordinated distribution of farm produce through their own wholesale link, and not under the control of the Farmers' Cooperative. The negative reaction from the largest retail chains must therefore be based on the reasoning that they will be unable to gain control of the farm-based share of the range of products.

As the market for farm products opens up, the threat that the grocery chains will choose foreign suppliers will become more real for the Farmers' Cooperative. At the same time, there are also some private actors (producers who are not members of the Farmers' Cooperative) in the Norwegian market, delivering farm products to the retailers. There are private suppliers in the meat, egg, poultry, and fruit and vegetable branches, but not for milk, where Norske Meierier (the Norwegian Dairies) have a monopoly. More extensive use of suppliers from outside the Farmers' Cooperative is a more genuine threat to the Farmers' Cooperative today, than grocery chains who chose foreign suppliers.

For their part, these private producers fear that collaboration within the Farmers' Cooperative will reduce their own power to negotiate with the grocery chains. A centralized organization or coordinated distribution within farm cooperatives will concentrate the farm-based share of the food trade. A common distribution system within the Farmers' Cooperative will cover the whole country, which will be an advantage in that the leading grocery chains are mainly nationwide and therefore demand nationwide coverage from those who supply them. The private share of meat production is mainly supplied by regional producers, for whom it may be difficult to satisfy the grocery chains' demand for nationwide distribution. The private meat companies therefore fear that coordinated distribution by the Farmers' Cooperative might accelerate the closing down or takeover of companies in the private sector. Because of this, they protested to the Bureau of Fair Trade[5] against the Farmers' Cooperatives' plans for coordinated distribution.

Both retailers and private producers are thus negative to the Farmer Cooperative's plans for a nationwide coordinated distribution of farm products. The message from most of the retailers is clear: 'If you won't play by our rules, we can find someone else to play for our team.' The question of how genuine this threat is depends on the degree to which the grocery chains are able to choose fresh produce suppliers from outside the Farmers' Cooperative.

There are, however, some retailers who are more positive to the Farmers' Cooperative's plans for distribution of fresh produce. Trond Lykke owns a regional food store chain in mid-Norway and his business is too small to manage its own distribution to any great extent, and coordinated distribution by the Farmer Cooperative would be beneficial to Lykke.

> Farming is running an industry – the industry of fresh produce. They ought to be allowed to protect their own competitive position, which means being able to coordinate their own distribution. The farmers can act as a corrective and alternative to private wholesalers. Both the grocers, the suppliers, and the customers ought to be happy about this. (Interview with Trond Lykke in *Dagens Næringsliv* 19.06.95)

There are already signs of private producers forming closer bonds with one of the large chain's distribution system. Some private meat producers have started wholesale distribution of their own products. For example, in July 1994, Rema-gross Kjeldsberg at Ski started distributing fresh produce from Kolbjørn Grobøl A/S at Gjøvik to approximately seventy REMA stores in the eastern region of the country (Blekesaune et al., 1995). These companies are required to follow the rules that the retailers have established: 'You produce and deliver to our warehouse, and we distribute.' This type of development opens opportunities for specialization of production within the private meat companies. It is then the wholesalers' task to put together a complete assortment of meat products for the stores, and the individual producer can then specialize within this range. Such alliances between the retail and wholesale links are regarded as a prerequisite for private meat producers to be able to continue. Given the increase in competition they will be faced with in the years to come, the farm producer has to protect his home market against foreign competition.

Conclusion

In this Chapter we have looked at what lies behind the struggle surrounding the distribution of food in Norway. We have studied the various interests tied to distribution among the farmer cooperatives and the grocery chains. The producers maintain that given the way wholesale distribution functions today, direct distribution is the cheapest or most cost-effective form of distribution. If wholesale distribution increases the costs of products, prices will rise and this will reduce demand and total turnover. Arguments against giving up direct distribution have been strongest from the suppliers of fresh produce, such as meat and dairy products. According to these suppliers, there are many factors that make direct distribution superior to wholesale distribution. The chains and the wholesalers maintain that direct distribution means extra costs for the consumer, and that this form of distribution is neither rational nor efficient in relation to the individual stores' receiving routines.

The largest grocery chains' latest move in this game is joint cooperation between the REMA group, the Hakon group, the Consumer Cooperative (NKL) and Norwegian Retailers as a major effort in wholesale fresh produce. They want to prevent the Farmers' Cooperative's plans from being realized at any cost. The private wholesalers (basically owned and controlled by the retailers) have on their side included the farm industry's transport in its own plans for growth a long time ago.

During our study, we have identified some issues of great interest for further studies of the development of the Farmers' Cooperatives. One of these issues deals with farmers' solidarity. The Cooperative as such could be seen as a kind of 'solidarity movement,' and a central question for future research is whether the new generations of farmers will continue to cooperate or not. Even if everybody benefits from cooperating, the benefits for each individual farmer could be better if he (or she) sold his (or her) products outside the cooperative, at least in the short run. And in the long run it will be a benefit providing that other producers are acting pragmatically (Olson, 1965). Individual rationality based on egoism results in abandoning the strategy of collaboration, even if the final result then will be less favourable than if all had collaborated.

The overwhelming proportion of farmers involved in the cooperatives makes the Norwegian case distinctive. As we see it, the question of whether or not the Farmers' Cooperative has enough weight to create a genuine countervailing power in their confrontation with the large grocery chains will depend on the degree to which the Farmers' Cooperative is able to make use of its monopoly on supplying milk. The grocery chains do not seem to be interested in distributing milk (*Bondebladet* 21-06-95). The distribution of milk must take place relatively often, and this should make it possible for the Farmers' Cooperatives to include other farm products in the same truck.

Another important task for the Farmers' Cooperative in the future, will probably be to 'educate' Norwegian consumers to prefer farm products produced in Norway. The national food culture, taste and traditions will be important factors when it comes to the strengthening of Norwegian produced farm products in competition with cheaper products from other countries. Seen in this context, a common distribution system for fresh produce could be a central agent.

The challenges that now face primary producers, with dominating chains and increased internationalization, demand new ways of thinking in order to be able to fulfil the idea of cooperation. The Farmers' Cooperative has several strategies to choose from in order to strengthen its own position in its confrontation with the powerful grocery chains. Establishing a nation-wide coordinated distribution of farm products is one strategy. Another strategy for strengthening the competitive capabilities for Norwegian food both nationally and internationally is the Godt Norsk (Good Norwegian) campaign initiated by the Norwegian government. This campaign shows how federal funding still plays a key role in the Norwegian food retail sector. In many other countries, for instance in the UK, there have been considerable changes – mainly deregulation – in the functions of the state over the past decade. We can also say that state regulation of the grocery trade has decreased in Norway to a certain degree. There are, however, certain characteristic conditions for the Norwegian grocery trade which makes the Norwegian food retail sector distinctive, compared with other advanced industrialized countries. The Norwegian food retail sector is not privatized and commercialized to the same degree as elsewhere. The main components which makes the Norwegian situation distinctive is the strong position of the farmers' cooperatives in Norway, and the fact that the retail sector is still regulated to a certain extent by the state.

Notes

1. There are various forms of wholesale distribution in the food retail market, from independent wholesaler who is the middleman between the producer and the retailer, to the retailer in charge of distribution. Within the Consumer Cooperative system and the Hakon group we find, today, owner-integrated wholesale functions. REMA 1000 has entered mutually binding contracts for delivery with individual wholesalers, such as Norgesfrukt AS and R. Kjeldsberg AS.
2. Norwegian Retailers were not interested in participating in our study.
3. *Dagligvarehandelen* is an independent newspaper for the food retailer sector and is published by Trade Press A/S. *Fritt Kjøpmannskap* is published by Norges Colonialgrossisters Forbund (The Norwegian Association of Grocery Wholesalers).
4. In northern Norway the great geographical distances and the spread of the population has made integrated cooperation between the branches a natural strategy for

several decades (Hertzenberg, 1994). In addition to the practical experience with this type of cooperation in northern Norway, much experience has also been gained with the distribution of various farm products in shared trucks in parts of Nordmøre and Romsdal. Since 1991 these regions have also had a common collection terminal for certain types of agricultural products. There are also positive experiences with coordination of the distribution of farm products from a trial project in Vestfold (Hertzenberg, 1994).

5. The Bureau of Fair Trade is commissioned to see whether the stipulations in the Prisloven (Price act) are complied with.

References

Almås, R. and Ward N. (1994), 'The Norwegian Agro-Food Industry. Problems and Prospects,' in Almås, R. and Ward N. (eds), *Restructuring the Agro-Food System: Global Processes and National Responses,* Proceedings from an International conference on the restructuring of the agro-food system, organized by the Centre for Rural Research, Trondheim, May 2–4 1994.

Blekesaune, A., Nygård, B. and Storstad, O. (1994), *Ny struktur i matomsetningen. Konsekvenser av endringer i detaljomsetningen med mat i noen utvalgte europeiske land,* Report 2/1995, Centre for Rural Research: Trondheim.

Blom, S.E., Framnes, R. and Thjømøe, H.M. (1985), *Markedsføringsledelse*, Tano: Oslo

Flynn, A., Marsden, T. and Ward, N. (1994), 'Retailing, the Food System and the Regulatory State,' in Lowe, P, Marsden, T. and Whatmore, S. (eds): *Regulating Agriculture*, David Fulton Publishers: London.

Friedland, W. (1991), 'The Transnationalisation of Production and Consumption of Food and Fibre: Challenges for Rural Research,' in Almås, R. and With, N. (eds): *Rural Futures in an International World*, Centre for Rural Research: Trondheim.

Friedman, H. (1982), 'The Political Economy of Food: The Rise and Fall of the Post-War International Food Order,' *American Journal of Sociology*, Vol. 88, pp. 246–86

Galbraith, J.K. 1967, 'The Concept of Countervailing Power,' in Mallen, B. (ed.): *The Marketing Channel*, John Wiley & Sons Inc.: New York.

Goodman, D. and Redclift M. (1991), *Refashioning Nature. Food, Ecology and Culture,* Routledge: London.

Hertzenberg, O.G. (1994), 'Felles ferskvaredistribusjon: 450 mill. å spare,' *Landbrukssamvirket*, No. 2.

Hovi, J. and Rasch B.E. (1993), *Strategisk handling. Innføring i bruk av rasjonalitetsmodeller og spillteorier,* Universitetsforlaget: Oslo.

Jacobsen, E. and Dulsrud, A. (1994), *Kjedenes makt til forbrukerenes beste? Konsentrasjon og integrasjon i dagligvarehandelen med matvarer i Norge*, Statens institutt for forbruksforskning: Oslo, Report 4–1994

Miles, M.B. and Huberman A.M. (1984), *Qualitative Data Analysis,* Sage Publications: Newbury Park.

Munton, R. (1992), 'The Uneven Development of Capitalist Agriculture: The Repositioning of Agriculture within the Food System,' in Hoggart, K. (ed.), *Agricultural Change, Environment and Economy: Essays in honour of W.B. Morgan*, Mansell: London.

Olson, M. (1965), *The Logic of Collective Action. Public Goods and Theory of Groups,* Harvard University Press: Cambridge.

Rosengren, K.E. and Arvidson, B. (1983), *Sociologisk metodik.* Esselte Studium AB: Lund.

Wallace, I. (1992), 'International Restructuring of the Agri-Food Chain,' in Bowler, I., Bryant, C. and Nellis, M. (eds): *Contemporary Rural Systems in Transition:* Volume 1 – *Agriculture and Environment*, CAB International: Wallingford, Oxon.

Ward, N. (1990), 'A Preliminary Analysis of the UK Food Chain,' *Food Policy*, Vol. 13, pp. 439–441.

Wrigley, N. (1993), 'Retail Concentration and the Internationalisation of British Grocery Retailing,' in Bromley, R. and Thomas C. (eds), *Retail Change: Contemporary Issues*, University College London Press: London.

Part III

Agrarian Change and the Environment

5 The adoption of alternative farming enterprises in Ireland's Less Favoured Areas

Mary Cawley, Desmond A. Gillmor, Perpetua McDonagh

Introduction

At an international level, agriculture, like many other economic sectors, has been subject to major processes of restructuring in recent decades in response to the crisis of overproduction (Goodman and Redclift, 1989). Farm incomes from conventional agriculture have fallen and supplementary income sources have been sought in off-farm employment and in novel forms of non-conventional enterprises on the farm (Dalton and Wilson, 1989; Fuller, 1990; Kelly et al., 1992; MacKinnon et al., 1991; Shucksmith et al., 1989). Both the adoption of off-farm employment by resident farm members and on-farm diversification have been pursued in the past, most notably the former; supplementing income through the provision of farm accommodation also has a long history in many scenic rural areas (Halliday, 1989). The range of non-conventional activities and production in which farm households began to engage in response to the most recent income crisis was, however, unprecedented. So, too, was the scale of financial and other supports provided by the European Union (EU) and by national governments to encourage conversion from products that were in oversupply (Bowler, 1992).

As in many other countries of Western Europe, agriculture in the Republic of Ireland has become increasingly subject to processes of restructuring, not least because of enhanced integration within EU policy and market structures (Gillmor, 1987a). A growing dependence on price and subsidy supports from the Common Agricultural Policy (CAP) and an expansion of European markets for output followed Ireland's accession to membership of the Community in 1973 (Sheehy, 1984). Reform of the CAP to reduce overproduction, including price reductions, diminished subsidies and output controls, together with the influence of the Uruguay round of the General Agreement on Tariffs and Trade (GATT), exerted downward pressure on farm incomes during the 1980s (National Economic and Social Council, 1992). The income gap between various sectors of the farming population within Ireland also increased in the years after 1973, reflecting differing capacities to benefit from the higher profit activities, notably dairying (Walsh, 1991). Off-farm employment increased in importance as a means of supplementing inadequate farm incomes. From the late 1980s on, growing numbers of non-conventional enterprises were introduced on farms, as part of an adaptation to constraints associated with conventional enterprises. These included: the introduction of new crops and livestock, the products of which were not subject to quota restrictions; the provision of services on the farm, such as tourist accommodation; afforestation; and various types of farm-based processing and manufacturing (Cawley et al., 1995; Gillmor, 1992; Kelly et al., 1992; Willer and Gillmor, 1992). This trend was actively encouraged by the EU and by the national government through grant aid and other support measures. The need for supplementary sources of farm income has always been greatest in Ireland's western counties which merit Less Favoured Area (LFA) status. The capacity for adaptation to a rapidly changing external policy and market environment is also less in this area, and non-conventional enterprises have a potentially important role to play in the survival of many smaller family farms. The LFAs therefore form a particularly appropriate context for investigating the adoption of alternative enterprises.

For the purposes of the study reported here, alternative farming enterprises (AFEs) were interpreted as referring to types of crop and livestock production new to the region and not covered by price support; farm afforestation; adding value to food and non-food products on the farm by manufacturing, processing or direct marketing; and a wide range of on-farm service activities including recreational and accommodation activities. A total of sixty-two possible enterprises was identified in all (Cawley et al., 1995). The focus of attention was on non-traditional enterprises on the farm. Agricultural contracting, where farm capital and labour were involved in the off-farm activity, was also included. Other types of off-farm employment were excluded. In this context it is worth noting that, within a growing literature relating to the contemporary role of supplementary sources of farm income, the term 'pluriactivity' is usually used to include both employment off the farm and on-farm diversification (MacKinnon et al., 1991). Ilbery (1991) differentiates clearly between the two and provides a detailed classification of alternative farming enterprises, on which the present study is based.

Study method

A behavioural approach was used to investigate the adoption of AFEs within the context of a changing CAP. This approach is based on the premise that farm families make decisions relating to the farm business having regard to the human, physical and monetary capital of the farm, within the context of external influences (Bowler et al., 1993; Gillmor, 1987b; Ilbery, 1983). The external farm environment is conceptualized as the institutional complex associated with farm policy which includes education and training, market and capital structures, farmer representative organizations and broader farming interests. This external environment is of growing importance within an increasingly globalized agriculture.

A decision-making model was developed in the course of the research project in which farmers are viewed as being subject to stimuli for change (external and internal to the farm) which prompt them to undertake a search process to identify possible paths among which choices are made (Bowler et al., 1993). The model hypothesizes that search behaviour, triggered by the stimuli, is conditioned by two primary factors: the ability of the farm family to identify and assess the various options available to them and their use of the enabling services provided by elements external to the farm. In response to an assessment of the strategies identified, farm operators make decisions with regard to the farm enterprises to be pursued, which are envisaged as pathways. The three stages of stimulus, search and choice are traced in this Chapter through comparative study of AFE adopter and non-adopter families and farms. An analysis of the pathways selected is reported.

Ireland's LFAs, using the pre-1992 definition which was adopted for this study, cover some 58 per cent of the total area of the state (Figure 5.1). Investigation of the adoption process within this area was conducted through questionnaire surveying of a sample of farm operators in two counties: County Galway on the west coast and County Kerry in the southwest. One hundred farms were surveyed within each county, giving a total sample of two hundred farms. Sixty conventional farms were selected randomly for each county from the sample of farms chosen for the National Farm Survey (NFS) which is administered by Teagasc (the Agriculture and Food Development Authority): for the 11 per cent of these farms which were found to have AFEs, non-adopters were substituted from a random fall-back sample list. Additional AFE farms were selected to bring to forty the total of such farms in each county. The AFE farms provided a representative sample of different types of AFEs.

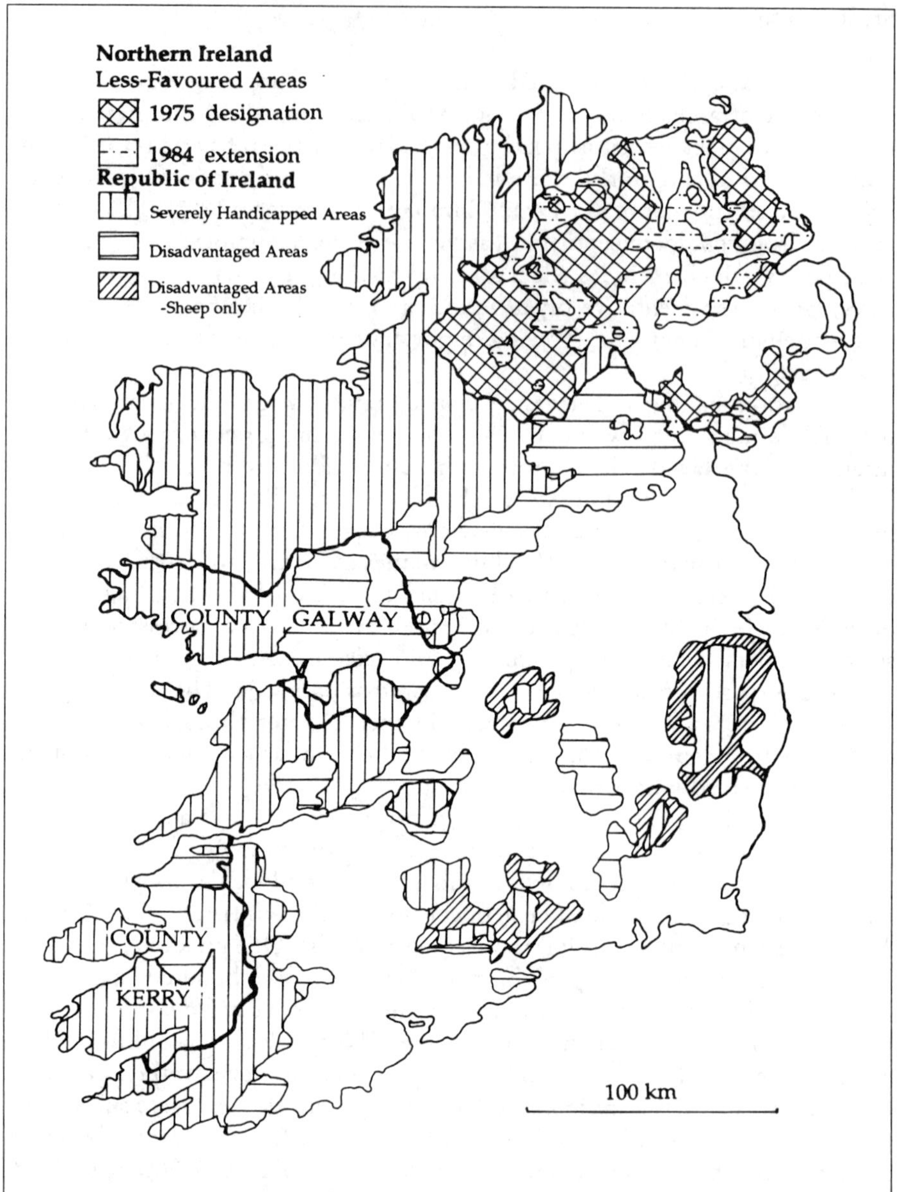

Figure 5.1: Less-favoured areas in Northern Ireland and the Republic of Ireland

Table 5.1: The range of AFEs surveyed

AFE description	No of enterprises adopted
Timber	24
Serviced accommodation	14
Sport horses	9
Self-catering accommodation	7
Deer, contracting	6 (each)
Organic crops, pony trekking	5 (each)
New vegetables, greyhounds, open farm	4 (each)
Free-range poultry, goats, farmhouse cafe, dairy processing	3 (each)
Golf course	2
Nursery, organic lifestock, snails, farm shop, freezer sales	1 (each)

The range of AFEs adopted

AFE farms were selected for study so as to ensure that the broad range of alternative farming enterprises currently practised in Ireland was represented proportionately in so far as that was possible. A wide diversity of AFEs was identified, of which forestry and serviced accommodation were of greatest importance (Table 5.1). Other AFEs which occurred on numbers of farms were sport horses, self-catering accommodation, deer, farm contracting and organic crops. A more extensive list of crop, livestock, processing, manufacturing and sales AFEs was represented on fewer farms. Some AFEs were developed in addition to mainstream enterprises and some were substituted for conventional enterprises in response to quota restrictions and falling prices. A limited number replaced the conventional farming system as, for example, where a farm was converted to a golf course. Fifty-eight farms had one AFE only and the remaining twenty-two had two or more AFEs. Farms with more than one AFE which provided serviced accommodation tended to have related activities, such as unserviced accommodation and pony trekking, and included 'open farms' with exotic animals and poultry on view. Some provided farmhouse teas. By contrast with accommodation, forestry tended to be associated with a range of unrelated activities including sport horses, mushrooms, accommodation and contracting. The majority of diversifiers (75 per cent) had introduced their AFEs since the late 1980s, although some serviced accommodation dated back to the 1960s. The recent introduction of many of the enterprises points to the growing pressure on farmers to seek non-conventional options in their quest to use farm resources effectively and to maintain incomes. Enhanced financial incentives to promote AFEs were also introduced by the EU and the national government from the late 1980s on.

Because of the importance that is being assigned to the promotion of AFEs by the national government and the EU as means of supplementing farm incomes and providing employment, it is pertinent to identify how people become aware of alternatives, and to establish the role of farm-family members in influencing the decision to adopt. A distinction was made in the study between external information sources and other influences which prompted the decision to diversify and the family member who initiated the decision to adopt a particular enterprise. The initial information which provided the stimulus to introduce the AFE was reported as coming primarily from organizations external to the farm, including media sources, Coillte (the state forestry company), tourism organizations, Teagasc and regional development organizations. General media information, including articles in the farming press relating to AFEs, was cited as being of particular importance. Other factors influencing the decision to diversify, which relate more to reasons for diversification than initial sources of awareness, included a personal or hobby interest, the existence of a market demand for a particular product, the availability of resources to invest in the enterprise, and the availability of grant assistance towards establishment costs (Cawley et al., 1995).

In relation to the actual decision to adopt a particular AFE, the farmer was reported as initiating the idea in over 50 per cent of cases for both a first and a second AFE, and in 40 per cent of cases for a third AFE. Where one AFE only was present, the spouse initiated the idea to diversify in 16.3 per cent of cases and in 12.5 per cent of cases the decision was reported as having been taken jointly by the farmer and spouse. Joint decision making with a spouse tended to be more common where the AFE involved substantial labour and managerial inputs by the, usually, female spouse and required liaison with the public as, for example, in the case of accommodation, open farms, farmhouse teas and cheese sales. The role of the spouse in the decision to adopt an AFE increased as the number of AFEs increased. On farms having two and three AFEs present, 31.8 per cent and 60 per cent, respectively, were proposed by the farmer's spouse.

The stimulus to adopt an AFE

The internal environment

The aspects of the internal farm environment which were viewed as having a possible influence on the adoption of an AFE were: (i) evidence of prior accumulation behaviour, which was defined according to selected criteria outlined below; and (ii) potential stress factors associated with the declining profitability of the conventional enterprises, demands for additional income to meet school or college fees and other costs at certain stages in the family cycle, the need to create employment for either family or non-family members, and a demand for an alternative lifestyle. Findings from earlier studies in other contexts suggest that a record of accumulation behaviour tends to predispose a farmer towards further accumulation (Shucksmith and Smith, 1991). The presence of children of college age in the household may create pressure on the existing farm income (Shucksmith, 1993). Declining profitability has been identified as influencing the search for alternative sources of income as has the need to create family employment (Gasson, 1988). Alternative lifestyle aspirations have been associated with the adoption of organic production in particular (Slee, 1987; Willer and Gillmor, 1992). The role of these various factors is assessed by comparing adopter farms and families with their non-adopter counterparts.

Accumulation behaviour and AFE adoption The factors that were investigated in relation to accumulation or entrepreneurial behaviour related to: the characteristics of the farming system practised, profit levels, the existence of forward planning, attitudes towards market opportunities and towards the availability of grant aid, previous experience of diversification, and identification of a specific marketing opportunity for an AFE. Analysis revealed that a relatively limited proportion of all farm operators in the study areas could be classified as accumulators, but diversifiers outweighed non-diversifiers in this regard (23.7 per cent of the former were so classified, versus 7.5 per cent of the latter). Accumulator farms with and without AFEs tended to share similar conventional enterprises. Intensive cattle and sheep farms with above-average gross margins per hectare dominated in both instances. The area of land owned by AFE farmers was less on average than by their non-AFE counterparts (34.9 ha as compared with 38.5 ha), although the former farmed relatively larger areas (55 ha on average in comparison with 49.5 ha), much of the additional land being in the form of commonage (land in which two or more farmers have grazing rights). The mean area rented by both groups of farmers was less than 5 ha. The dominance of farm ownership of less than 40 ha and the similarities in output mix reflect broad underlying agricultural patterns within the LFAs.

The incidence of farm profit and trends in profits during the five years 1988–1992 point to relatively higher profits being associated with adopter than with non-adopter farms. More AFE adopters than non-adopters reported low or nil farm profits, and similar proportions of both groups registered profits in excess of £10,000 per annum. Profits in the £5,001–£10,000 range were, however, much more characteristic of AFE farms than of their counterparts (Table 5.2). Increase in farm profits over the years 1988–1992 was also associated more closely with AFE than with non-AFE farms: 46.0 per cent of the former had increased their profits over the four years by comparison with 15.5 per cent of the latter. More of the former also reported a fall in profits over the same period of time which may be attributed, in part at least, to increased investment in the farm, including investment related to the introduction of an AFE.

Table 5.2: Financial stimuli in the internal environment for farm enterprise change

	AFE farms (%)	Non-AFE farms (%)
Gross farm profit		
Zero	11.3	2.5
>£5,000	25.0	46.7
£5,001–£10,000	40.1	29.2
>£10,000	21.3	20.8
No data	2.5	0.8
Trend in profits 1988–1992		
Increase	46.0	15.5
Decrease	26.4	16.3
No change	27.6	68.1
Levels of farm debt		
Zero amount	33.7	66.1
>£5,000	10.4	15.3
£5,001–£20,000	22.1	11.9
£20,001–£50,000	22.1	5.9
£50,001–£100,000	11.7	0.8
Trend in debts 1988–1992		
Increase	46.1	16.1
Decrease	26.3	15.8
No change	27.6	68.1

In terms of regular farm planning and awareness of markets, AFE farmers reported more progressive features. Some 41.3 per cent engaged in regular forward planning, by comparison with 15.8 per cent of non-AFE farmers. It is of interest in this context also that over one-third of the adopters had gained operational control of their farm within the previous ten years, by comparison with 19.2 per cent of non-adopters. This tends to suggest that land transfer, particularly to younger operators, may help to facilitate diversification. Awareness of markets and appreciation of their importance were also higher among adopters, whilst both groups considered grant aid to be important. Some 69 per cent of AFE adopters received grant aid on establishing the enterprise. AFEs established without grant assistance consisted in the main of bed and breakfast activities where aid was not available for this particular type of diversification or was not available at a particular period in time. Only eighteen respondents had prior experience of an AFE before its introduction and only twenty were responding to a specific marketing opportunity. The low levels of prior experience reinforce the importance that was attached to external sources of information in arriving at the decision to diversify, as noted earlier.

The diversified farms in the study areas shared features in common with conventional farms in terms of size and the enterprises practised. At the same time, larger farm businesses were associated with the adoption of AFEs, and the intensity of land use may have influenced the financial ability of a farmer to diversify. Diversification was viewed as being a risk because of a fear of the unknown and the untested, and also because of the associated capital requirements. For diversifiers, the perception of risk was, however, outweighed by the desire to generate additional income. Economies of scale no longer provided adequate opportunities, given restrictions on output. There was no evidence that diversifiers employed new technology to any greater extent than their non-diversifying counterparts.

Stress factors and AFE adoption Two main potential sources of financial stress were examined, namely (i) the need to service farm debt and (ii) the dependency ratio in the household as an indicator of a demand for higher income at particular stages in the family cycle. A majority of both types of respondents expressed the view that access to personal finance and borrowed finance for investment was important for the development of an AFE. The evidence in relation to debt levels reveals, however, that borrowing is more characteristic of AFE farms. One-third of the AFE group and two-thirds of the non-AFE group had no borrowings (Table 5.2). The amounts borrowed were also considerably higher among the former than among the latter: some 11.7 per cent of the former had borrowings of £50,001–£100,000 by comparison with less than 1 per cent of the latter; and 22.1 per cent and 5.9 per cent, respectively, owed £20,001–£50,000. Over half of the AFE group had borrowed to cover the establishment and maintenance of the AFE and 46 per cent reported increased debts between 1988 and 1992. Debt level seems to be a result of adoption rather than being a stimulus for it, a finding that conforms with that of Ilbery and Bowler (1993) relating to farm diversification in England and Wales. The highest levels of debt were associated with farms that had introduced on-farm services (notably accommodation enterprises) where investment in buildings, furnishings and recreational facilities was necessary, and with new livestock enterprises such as deer where investment was required in special fencing, the purchase of the breeding stock and the modification of farm buildings. It is of some interest to note that afforestation, an important user of land resources, was not associated with large-scale borrowing: instead, it appears that underutilized land was brought into use for this purpose and that grant payments covered much of the investment costs. Short term borrowing may have been necessary whilst awaiting the receipt of a grant.

Table 5.3: Search behaviour: family life cycle stimuli for farm enterprise change

Life cycle stage	AFE farms (%)	Non-AFE farms (%)
Farmer >45, no others resident	3.8	24.2
Farmer >45, spouse only resident	6.3	7.5
Farmer with/without spouse		
All resident children >16	13.8	16.7
At least one child >6	35.0	26.7
Farmer with spouse		
At least one child <6	18.8	6.6
Farmer <45, spouse resident, no others resident	8.8	0.8
Farmer <45, no spouse, no children resident	7.5	17.5
Not classified	6.0	0.0

The relationship between certain stages in the family cycle and the quest for supplementary sources of farm family income is a recognized phenomenon (Gasson et al., 1988). Households were classified according to composition, the age of the farmer and of any children who were present (Table 5.3). Over 80 per cent of the total were single family households. In the remainder, adult relatives of the nuclear family were present. The average age of the farmer was lower on AFE (47 years) than on non-AFE farms (53 years) but relatively low proportions of farmers overall were under 45 years. This age profile contrasts with findings by Kelly et al. (1992) from AFE farms in County Wicklow in eastern Ireland where 60 per cent of operators who had adopted alternatives were aged less than 40 years. Almost 42 per cent of non-diversifiers in the LFAs lived alone, by comparison with 11.3 per cent of diversifiers. There is limited evidence that the presence of school-going children is associated with the adoption of an AFE. Households with at least one child under 16 years were most likely to have diversified, especially if the child was over 6 years of age. However, 40.2 per cent of AFE adopters had no children under 16 years and there was no apparent relationship between the number of children in the household and diversification. A slightly higher proportion of AFE than of non-AFE households had children in continuing education which would have created some demand for college fees. Unlike findings by Ilbery and Bowler (1993) from Britain, the research did not find that the existence of a farm successor acted as a stimulus for diversification as a means of ensuring the longer-term viability of the farm business. Farm operators who had diversified, in fact, expressed greater uncertainty about the future of their farm units as family businesses than did non-diversifiers. A reluctance on the part of other family members (sons and daughters) to continue in the agricultural tradition in which they were reared, or the adoption of other careers on their part, may be critical in this regard, as opposed to indecision on the part of the farm operator or the absence of a potential successor. The perceived declining viability of the farm holding, which may have prompted the diversification, is pertinent also to the career aspirations of farm family members.

A desire to create employment for either family members or for hired labour, which was hypothesized as being a possible stress factor, was not found to be significantly associated with diversification. Only eight respondents who had introduced an AFE cited the need to create employment for family members as one of their reasons for its introduction and only two respondents stressed the creation of non-family employment as the motivating factor. Over 80 per cent of both AFE and non-AFE respondents felt, however, that the availability of family labour would be important or very important for setting up an AFE. Relatively lower proportions considered the availability of hired skilled labour to be important, reflecting the low levels of hired labour within the LFAs and also the dominance of small farms and types of enterprises which have limited demands for labour. Labour availability was highest on AFE farms with some 29 per cent having in excess of one family work unit by comparison with 17.5 per cent of non-AFE farms. Labour was underutilized on both types of farm, thereby suggesting that per se it does not act as a strong stimulus for diversification. An aspiration for an alternative lifestyle, the third stress factor to be examined, was rarely mentioned as a reason for diversifying, except by farmers involved in organic production. Almost 70 per cent of AFE respondents claimed, however, that the introduction of the enterprise had increased the quality of life for their families.

The primary reasons cited for not adopting an AFE reveal that preferences for traditional farming practices, insufficient capital, lack of knowledge, age and risk aversion were the main barriers involved. These are indicative of the general absence of accumulation tendencies among conventional farm operators.

The external environment

Farming takes place within an increasingly complex external environment which involves not only policy and marketing agencies but contacts with financial institutions and with a growing number of information sources (Gillmor, 1992; Marsden et al., 1986; Moran and Cockling, 1989). In Ireland, the farm advisory service has also been allocated new roles in promoting and supporting AFEs. Contact with a diversity of external agencies is therefore a necessary part of contemporary farming. Both diversifiers and non-diversifiers recognized the importance of agents working outside the farm unit. Some 90 per cent of both groups of farmers considered the availability of government financial assistance, training courses, marketing organizations and advisory and information services to be either very important or important for establishing an AFE. AFE adopters were found, however, to be more likely to have contact with external organizations than were their counterparts: 54.2 per cent of AFE farms and 45.8 per cent of non-AFE farms liaised with external agents. The contrast in the incidence of contact with external agencies was most marked in relation to the farm advisory service: 63.8 per cent of adopters availed themselves of the service by comparison with 35 per cent of non-adopters. AFE adopters appeared to be better networked and had greater accessibility to information, but further research is required in order to justify the hypothesis that external contacts actually stimulated AFE development. Farmers' awareness of other diversifiers in their immediate area of residence was relatively low, underlining a dependence on agencies and on the media as sources of information pertaining to AFEs. The relative recency of the introduction of AFEs, the limited experience of their operation and the restricted numbers of farmers involved place certain limits on the opportunities for the exchange of information between farmers.

Analysis of the stimuli for AFE adoption highlights the complexity of the decision-making process. It is very difficult to conclude that there is evidence to suggest that either entrepreneurial behaviour or survival strategies, of themselves, can explain what motivates a farmer to adopt an AFE. It would appear that a combination of both sets of factors is involved in many cases in Ireland's LFAs. A majority of the farms in the study areas were not, by virtue of location and resources, associated with accumulation tendencies. A relatively small proportion of all farmers could be classified as accumulators. The primary stimulus for AFE adoption for a majority of diversifiers was the maintenance or increase of family income. Because of the structural and business characteristics of the farms involved (relatively small extensive operations with low levels of efficiency and profit), the introduction of an AFE as a method of increasing returns is best interpreted as a survival rather than an accumulation strategy. At the same time, certain variables emerge from the analysis as serving to distinguish between diversifiers and non-diversifiers. These include important characteristics of the farm family, namely the ability of decision makers to identify opportunities presented by the external farm environment, including grant aid and market outlets, their abilities to manage their holdings and to secure credit finance, and their personal interest in the development of the farm business and the AFE. This combination of characteristics reflects a move towards entrepreneurial behaviour. AFE adopters also included younger farmers who had gained operational control of their farms during the ten years preceding the survey. Taken together, the findings foster the suggestion that stress-related factors are the primary stimuli for farm enterprise diversification but that the ability to respond to these stimuli and to overcome the problems they represent depend heavily on capacity for entrepreneurial behaviour.

Search behaviour for an AFE

Internal factors

Farm decision making takes place within an uncertain environment and it is recognized that in this context 'satisficing' rather than 'maximizing' choices tend to be made 'because of the incomplete nature of the information (available) and the complexity of the decision making' (Anosike and Coughenour, 1990, p. 3). Factors which were anticipated to contribute to optimum choices being made within a limiting environment were prior experience of farm adjustment, the involvement of family members in farm planning and levels of education and training. All of these factors may be viewed as constituting aspects of the human capital of the farm business.

Farmers who had adopted an AFE were found to have pursued adjustment strategies on their farms in the past to a greater extent than non-diversifiers: 93.7 per cent of the former reported some structural change on their farms between 1988 and 1992, by comparison with 69.2 per cent of the latter. Effective farm decision making is enhanced through use of the available knowledge and skill among the farm family, particularly when a spouse and family members may have higher formal levels of education than the farmer (Evans and

Ilbery, 1993). Recognition of the potential contributions of the larger stock of human capital on the farm was more common on diversified than on non-diversified farms. The farmer was the sole decision maker on 57.5 per cent of conventional farms whilst decisions were made by at least two members of the household (usually the farmer and his spouse) on 62.5 per cent of diversified farms. Involvement of family members in book-keeping and enterprise management was more characteristic of the latter also, reflecting the presence of expertise in this regard. Some of the conventional farms employed the services of accountants or of other financial advisors.

Table 5.4: Search behaviour: educational background of primary decision makers

	AFE farms		Non-AFE farms	
	Farmer (%)	Spouse (%)	Farmer (%)	Spouse (%)
Education				
No second level	29.1	14.3	64.2	36.6
Second level to age 14	2.5	2.9	2.5	1.4
Second level to age 16	31.6	20.0	21.7	18.3
Second level to age 18	22.8	35.7	9.2	39.4
Post second level	13.9	27.1	2.5	4.2
Training				
No training	45.2	38.1	75.8	81.0
Part-time course completed	23.2	15.9	18.3	7.9
Basic training	21.9	23.8	1.7	1.6
Agricultural degree	1.4	3.2	0.8	1.6
Other degree	8.2	19.0	3.3	7.9

In so far as the search process associated with farm adaptation in response to changing policy and market conditions requires an ability to identify and use information sources effectively, levels of education and training of the farmer and his/her spouse are undoubtedly of considerable importance. Educational capital emerged as firmly favouring AFE adopters: 70 per cent had some second level education and 54.8 per cent had some formal training (Table 5.4). Their spouses also had higher levels of education and training than their conventional counterparts, although not necessarily relating directly to agriculture: teaching and nursing qualifications tended to be most widely represented. Nevertheless, formal qualifications in the form of the school Leaving Certificate or a degree were held by only one-third and less than one-tenth of diversifiers, respectively. Over 72 per cent of diversifiers had no specific training relating to the alternative enterprise and 77.5 per cent had no previous experience of the enterprise. The limited proportions of farmers who had completed the second-level cycle undoubtedly reflects age characteristics, in part. Free access to secondary education was not introduced in Ireland until the late 1960s and prior to that time many sons of farmers on small farm holdings would have tended to complete their education at the age of 16 years, frequently having followed a technical course rather than an academic one.

Only 16.3 per cent of adopters had conducted a technical feasibility study in advance of introducing the AFE and only 13.7 per cent had investigated the potential market for the products of their AFEs. Of the eighty families with one or more AFEs who were interviewed, thirty-three had considered and decided against introducing another AFE, indicating that assessment of various types of alternative enterprises forms part of the search process. These included a wide range of enterprises. The main reason for non-adoption was insufficient finance.

External factors

Higher levels of contact with external institutions, and with the farm advisory service in particular, have already been recorded for AFE adopters compared with their counterparts. Virtually all of the AFE farmers liaised with one or more external institution, but only 55 per cent of the conventional farmers did so and with fewer institutions. The majority of contacts were with agricultural, financial and regulatory institutions. Diversifiers also had links with forestry and tourism organizations. The initiation of the overwhelming majority of contacts related to AFE development followed closely by contacts relating to the general development of the conventional farm business. The eighty AFE adopters initiated 177 contacts relating to the introduction of an AFE and ninety-two relating to an existing AFE. Approximately one-third of all surveyed farmers contacted institutions in the context of grant aid. The development of an existing conventional enterprise and general farm management advice were the main reasons for the initiation of contacts by non-diversifiers.

The results suggest that the AFE farm operators and their families possess human capital assets which assist in the search for adaptation strategies. These assets combined with joint decision making and superior management practices contribute to the successful identification and introduction of appropriate AFEs. It is important that younger, well-educated members of families become involved in this process, if the potential for the survival of family farms is to be increased. It must be recognized, however, that their involvement does not necessarily secure their participation in the business in the future.

Choice of paths of farm business development

The behavioural approach hypothesizes that having gone through the search procedure, farmers choose certain alternative courses of action or pathways of farm business activity. Six main paths of farm business development, identified by Bowler et al. (1993) on the basis of a typology devised by Whatmore et al. (1987), were used in this study. These are the traditional pathway, adoption of an agricultural AFE, adoption of a non-agricultural AFE, adoption of another gainful activity (OGA), industrialization of the farm business and winding down of the farm business. With the exception of the traditional pathway which, by definition, cannot occur with any other choice, the paths are not mutually exclusive: they classify business activity, not farms, so that a farm may belong to more than one pathway.

Table 5.5: Combinations of adjustment pathways

Pathways	Number of farms	%
Agricultural AFE	17	8.5
Non-agricultural AFE	12	6.0
OGA	24	12.0
Winding down	13	6.5
Traditional	79	39.5
Sub total	*145*	*72.5*
Agricultural AFE, OGA	23	11.5
Non-agricultural AFE, OGA	8	4.0
Agricultural AFE, Non-agricultural AFE	3	1.5
Agricultural AFE, Non-agricultural AFE, OGA	3	1.5
Agricultural AFE, Winding down	4	2.0
Sub total	*41*	*20.5*
Other combinations (of which 7 include AFEs)	14	7.0
Total	*200*	*100.0*

It is of interest to note that one-third of the farmers had adopted a single path diversification choice: 8.5 per cent adopted an agricultural AFE only, 6 per cent adopted a non-agricultural AFE only, 12 per cent sought off-farm employment only, and 6.5 per cent decided to disengage (Table 5.5). Taken with the traditional farms this means that 72.5 per cent of all farms selected one pathway only. The high proportion of farmers registering one pathway only is influenced in part by the fact that the non-AFE farms formed 60 per cent of the total sample and that the AFE pathway, by definition, was not a choice for them. Nevertheless, a potential existed for this group to register a combination of employment off the farm with either the industrial pathway or a 'winding down' of the agricultural enterprise. The fact that two-thirds of the non-AFE farms chose to pursue the traditional pathway is therefore of considerable note. The preference for adopting individual adjustment strategies rather than introducing multiple changes concurrently reflects a tendency towards risk aversion. Combined business strategies were introduced on fifty-five farms (27.5 per cent of the total). The most popular combinations were the association of an agricultural or non-agricultural AFE with the adoption of off-farm employment by a family member, usually by the farmer and, or, the farmer's spouse. The importance of other gainful activities (OGAs) as an adaptation pathway choice either alone or in combination with other choices is notable. It reflects the established role of off-farm employment in supplementing incomes on small farms in western Ireland. It also illustrates that the adoption of an AFE is being considered not only within the context of the conventional enterprise but with regard to other uses of labour and other sources of income.

Conclusion

Overproduction in conventional agriculture, falling prices and reduced state supports have led to a quest for non-traditional methods of resource use and income support in recent years. Alternative farming enterprises are being assigned a new role in this respect. The research reported here, which forms part of a larger study conducted in five EU countries, investigated the adoption of AFEs within Ireland's LFAs. A behavioural decision-making approach was used to analyse the adoption process within the context of Counties Galway and Kerry. The results of the research complement information which is emerging in Ireland relating to diversification and tends to confirm some findings of recent research in Britain.

A wide range of AFEs has been introduced in Ireland since the late 1980s in particular, although some forms of accommodation date to the 1960s. The most popular AFEs include farmhouse accommodation, forestry, new livestock and new crop enterprises. Both formal agencies and informal media sources constitute important sources of information on which decisions are made relating to adoption. The principal factors identified as contributing to the adoption decision include pressure on farm incomes arising from processes of restructuring, the availability of underutilized resources on the farm, an interest in a particular enterprise and a desire to benefit from a specific market opportunity. Whilst the provision of employment for either family or non-family members did not emerge as a major reason for adopting an AFE, the availability of underutilized labour on the farm may have facilitated the adoption and development of a non-conventional enterprise. Financial stress associated with specific farm debt burdens or with family educational costs did not emerge as particularly strong reasons for diversification: debt appears to have been more an outcome than a stimulus in this regard. Previous experience of farm system adaptation (although not necessarily the introduction of an AFE), involvement of family members in decision making and higher educational and training levels of the farmer and spouse were conducive to the search for appropriate AFEs.

Adopter farm households possessed other characteristics which distinguished them from their non-adopter counterparts. These included abilities to identify market opportunities, to secure credit finance as reflected in borrowings, and high levels of commitment to the development of both the conventional and the alternative farming enterprises. They were more likely to engage in regular business planning than were non-adopters and to have higher profit levels. Diversifiers maintained a wider range of institutional contacts than their non-diversifying counterparts and these contacts were closely associated with the introduction of an alternative enterprise. The personal and family attributes of AFE adopters and certain aspects of their behaviour point towards the presence of entrepreneurial characteristics and accumulation tendencies. Yet, like many of their non-adopter counterparts, adopter farms possessed features that conformed more closely to a survivalist model. These include the dominance of relatively small farm size and of low intensity sheep and beef enterprises among both groups. The ability to overcome such internal/external constraints as well as those imposed by restrictions on output depended on capacity for entrepreneurial behaviour.

A series of six possible pathways of farm business development was used to classify the farms in the study. Single pathways dominated overall with the adoption of an agricultural AFE or of a non-agricultural AFE occurring in 8.5 per cent and 6 per cent of cases, respectively. The adoption of another gainful activity, namely off-farm employment, was a more popular choice than either type of AFE individually and was also a popular combination with an agricultural AFE. The pathway analysis illustrates that whilst on-farm diversification has certainly increased in importance since the late 1980s, in response to both income pressure and the availability of incentives, it is not the only adjustment strategy being pursued in Ireland's LFAs. Off-farm employment is of considerable importance, either alone or in combination with other options, including

AFEs. It is clear that agricultural development in western Ireland, and the further promotion of AFEs in particular, needs to recognize and to accommodate the continued role of opportunities off the farm within the small farm economy.

Acknowledgement

This publication arises out of a programme of collaborative research by the following: Departments of Geography at the Universities of Leicester, Caen and Trinity College Dublin; Scottish Agricultural College (Aberdeen); CEMAGREF (Clermont Ferrand); Teagasc (Dublin) and Department of Economics at the University of Patras. The research project was funded under the European Union CAMAR Programme. The publication uses a farm questionnaire methodology developed by the Leicester and Caen laboratories during the collaborative research programme.

References

Anosike, N. and Coughenour, C. (1990), 'The Socio-Economic Basis of Farm Enterprise Diversification Decisions,' *Rural Sociology*, Vol. 55, No. 1, pp. 1–24.

Bowler, I. (1992), *The Geography of Agriculture in Developed Market Economies*, Longman: London.

Bowler, I. et al. (1993), *Conceptualising Farm Business Change and the Development of Alternative Farm Enterprises*, Working paper 3, CAMAR project on Alternative Farming Enterprises in the Lagging Regions of the EC, Department of Geography, University of Leicester: Leicester.

Cawley, M. et al. (1995), *Farm Diversification: Studies Relating to the West of Ireland*, Teagasc: Dublin.

Dalton, G.E. and Wilson, C. (1989), *Farm Diversification in Scotland*, SAC Economic Report No. 12, Scottish Agricultural College: Aberdeen.

Evans, N.J. and Ilbery, B.W. (1993), 'The Pluriactivity, Part-Time Farming, and Farm Diversification Debate,' *Environment and Planning A*, Vol. 25, pp. 945–59.

Fuller, A. (1990), 'From Part-Time Farming to Pluriactivity: A Decade of Change in Rural Europe,' *Journal of Rural Studies*, Vol. 6, No. 4, pp. 361–71.

Gasson, R. (1988), 'Farm Diversification and Rural Development', *Journal of Agricultural Economics*, Vol. 39, No. 2, pp. 175–82.

Gasson, R. et al. (1988), 'The Farm as a Family Business: A Review,' *Journal of Agricultural Economics,* Vol. 39, No. 1, pp. 1–44.

Gillmor, D.A. (1987a), 'Agricultural Development,' in Carter, R.W.G. and Parker, A.J. (eds), *Ireland: Contemporary Perspectives on a Land and its People*, Routledge: London.

Gillmor, D.A. (1987b), 'Behavioural Studies in Agriculture: Goals, Values and Enterprise Choice,' *Irish Journal of Agricultural Economics and Rural Sociology*, Vol. 9, pp. 19–34.

Gillmor, D.A. (1992), 'The Upsurge in Private Afforestation in the Republic of Ireland,' *Irish Geography*, Vol. 25, No. 1, pp. 89–97.

Goodman, D. and Redclift, M. (eds) (1989), *The International Farm Crisis*, Macmillan: London.

Halliday, J.E. (1989), 'Attitudes Towards Farm Diversification: Results from a Survey of Devon Farms,' *Journal of Agricultural Economics*, Vol. 40, No. 1, pp. 93–100.

Ilbery, B.W. (1983), 'Goals and Values of Hop Farmers,' *Transactions of the Institute of British Geographers, New Series*, Vol. 8, No. 3, pp. 329–41.

Ilbery, B.W. (1991), 'Farm Diversification as an Adjustment Strategy on the Urban Fringe of the West Midlands,' *Journal of Rural Studies*, Vol. 7, No. 3, pp. 207–18.

Ilbery, B.W. and Bowler, I.R. (1993), 'The Farm Diversification Grant Scheme: Adoption and Nonadoption in England and Wales,' *Environment and Planning C: Government and Policy*, Vol. 11, pp. 161–70.

Kelly, C.E., Ilbery, B.W. and Gillmor, D.A. (1992), 'Farm Diversification in Ireland: Evidence from County Wicklow,' *Irish Geography*, Vol. 25, No. 1, pp. 23–32.

MacKinnon, N. et. al. (1991), 'Pluriactivity, Structural Change and Farm Household Vulnerability in Western Europe,' *Sociologia Ruralis*, Vol. 31, No. 1, pp. 58–71.

Marsden, T.K. et al. (1986), 'The Restructuring Process and Economic Centrality in Capitalist Agriculture,' *Journal of Rural Studies*, Vol. 2, pp. 271–80.

Moran, W. and Cockling, C. (1989), 'The State and Rural Systems,' *Journal of Rural Studies*, Vol. 5, No. 4, pp. 313–24.

National Economic and Social Council (1992), *The Impact of Reform of the Common Agricultural Policy*, Government Publications: Dublin.

Sheehy, S.J. (1984), 'The Common Agricultural Policy and Ireland,' in Drudy, P.J. and McAleese, D. (eds), *Irish Studies 3: Ireland and the European Community*, Cambridge University Press: Cambridge.

Shucksmith, M. (1993), 'Farm Household Behaviour and the Transition to Postproductivism,' *Journal of Agricultural Economics*, Vol. 44, No. 3, pp. 466–78.

Shucksmith, M. et al. (1989), 'Pluriactivity, Farm Structures and Rural Change,' *Journal of Agricultural Economics*, Vol. 40, No. 3, pp. 345–60.

Shucksmith, M., and Smith, R.J. (1991), 'Farm Household Strategies and Pluriactivity in Upland Scotland,' *Journal of Agricultural Economics*, Vol. 42, No. 3, pp. 340–53.

Slee, B. (1989), *Alternative Farm Enterprises*, Farming Press: Ipswich.

Walsh, J.A. (1991), 'A Regional Analysis of Enterprise Substitution in Irish Agriculture in the Context of a Changing CAP,' *Irish Geography*, Vol. 24, No. 1, pp. 10–23.

Whatmore, S. et al. (1987), 'Interpreting a Relational Typology of Farm Business in Southern England,' *Sociologia Ruralis*, Vol. 27, No. 2/3, pp. 103–22.

Willer, H. and Gillmor, D.A. (1992), 'Organic Agriculture in the Republic of Ireland,' *Irish Geography*, Vol. 25, No. 2, pp. 149–59.

6 Unprotected land. Greek land-use policies in a period of farming contraction[1]

Leonidas Louloudis and Evi Arahoviti

Introduction

The CAP reform of 1992 was the most ambitious attempt so far to tackle the problem of agricultural surpluses in the EU, by limiting the high levels of price support for farm products and compensating farmers with a system of direct payments (Poux et al., 1995). This gradual but clear move away from market subsidies as a means of supporting farm incomes may reduce the attractiveness of farming in high productivity areas, as well as increasing the pressure for more fundamental changes in land use, such as abandonment of agriculture, afforestation, conversion to other use (urbanization, habitat conservation etc.). Thus land withdrawal from agricultural production and land abandonment are likely to increase in the future; the former due more to active policy measures, such as set-aside, than to a sharp cut in prices, and the latter due more to lack of favourable natural, economic and social conditions for farming in marginal areas (Baldock et al., 1993).

Yet the 1992 reform and with it the main proposals for restructuring market regimes (especially agro-environmental regulation – Reg. 2078/92 – from the so-called 'accompanying measures'), bring forward the issue of alternative management of farmland. However, in countries such as Greece, where even agricultural land of high productivity is already under strong pressure to change its use (e.g., to urban use), it is, we believe, a crucial issue whether the existing legal framework and regulatory practices effectively protect and control changes of land use. Obviously, under these circumstances, the proposed CAP restructuring measures, which could only indirectly be a useful policy tool for the conservation of agricultural land, are of marginal utility.

In Greece there is a multi-layered complex of laws, regulations and directives, aiming at rational exploitation and conservation of land, as well as controlling land-use changes. In practice though, as will be discussed below, this complicated system is either not applied at all or applied inadequately. A number of other factors exercise decisive influence on the regulation of land use in the country, the most important being specific land property characteristics and land market requirements.

These factors are discussed at various points later in this Chapter but it is worth noting here that the land property regime in Greece has one feature which is unique among other EU member states: land property rights include, apart from exploitation rights resulting from land characteristics and qualities, the right to build. Although the present Constitution does not establish it explicitly, this right has been established in practice and its abolition is considered particularly burdensome (Bieratos et al., 1995).

The claim and exercise of such a right is extremely pressing, given the dominance of medium- to small-scale land holdings and the high levels of rent. The existing legal framework has been based upon the possibility of building anywhere in the countryside, with only few exceptions.[2] Agricultural land is not an exception to that general rule. The continuous contraction of the farmed area due to the expansion of non-agricultural activities (industrial, tourist, residential, etc.) on agricultural plots, is common throughout the country. Even agricultural land of high productivity or land that in certain areas (e.g., islands) is considered important for the local economy, is threatened by uncontrolled urban expansion (Greek Agricultural Ministry, 1992). Indeed, it is the more productive land that is usually withdrawn from agriculture and not the land of marginal productivity, towards which non-agricultural activities are theoretically oriented, according to restrictions of the existing legal framework for the protection of agricultural land (Act 1337/1983). Therefore, land use is in practice determined by each landowner or investor and his/her individual aspirations.

Institutional protection of agricultural land

The new Constitution of 1976 includes a series of articles including the general framework for drawing up measures regulating land use under strict environmental conditions (art. 24, 17, 18, 106, 117). Some of these regulations for land-use planning and environmental protection are included in urban development (Act 1337/83) and environmental (Acts 360/76 and 1650/86) legislation in accordance with the above constitutional mandates.

Act 1337/83 for the 'Expansion of urban plans, urban development and relevant issues' allows for the first time, a provision for the protection of agricultural land, giving priority to highly productive areas. Thus, according to art. 1 par. 5, schemes for urban and industrial development on highly productive agricultural land under the Master Plan, are not allowed. Highly productive agricultural land is defined not only by its productive potential, but also because it belongs to an historical landscape (e.g., Amfissa's olive grove near Delphi), it is known for its traditional cultivations (e.g., 'masticha' gum trees on Chios island), or its ecological characteristics are of specific natural value (peat soils on the Philipi wetlands), etc. As an exception, expansion of urban plans in the above mentioned areas may be allowed only for permanent residence, when the specific towns or villages concerned are surrounded by such areas. If it is confirmed that there is no alternative solution, then expansion on highly productive agricultural land is permitted under specific terms and restrictions.

As far as the rest of agricultural land is concerned, it is possible to change use under the Master Plan exclusively on marginal lands. Finally, Act 1337/83 establishes the Urban Development Controlled Zones (UDCZs) which have been a basic tool for the regulation of land uses. UDCZs represent an attempt to control land use by imposing restrictions on building and sub-dividing plots, mainly for peri-urban areas but also for rural areas suffering more pressure (Greek Agricultural Ministry, 1983). However, although this policy has been applied for more than ten years, only a very small portion of UDCZs (fifty-five out of hundreds of UDCZs that had been planned in the period 1983–1993), has finally been approved and enacted by Decrees. The application of UDCZs in practice has encountered many problems of a technical, legal and socio-political nature which in many cases defeat the main purpose of spatial development regulation for the particular area. In short, the main weakness of this legal framework (i.e., Act 1337/83), is that protection of agricultural land is only achieved indirectly. In practice, within the above-mentioned spatial development framework, the protection of agricultural land is considered of secondary importance and is eventually secured only to the extent that the achievement of the main target permits.[3]

A more recent attempt at extensive land use planning has been based on the preparation of Special Land Use Plans (SLUPs) which finally led to the determination of UDCZs.[4] One basic characteristic of the SLUPs is that they refer to a sub-prefectural level.[5] This small-scale spatial development planning however, can by no means substitute for spatial development plans of a larger scale. Still, this does not mean that SLUPs are insignificant, for they play a specific role that cannot be performed by regional or national spatial development plans.

The designation of land use and building regulations in the countryside is the main outcome of spatial development and it can be accomplished only through a detailed and specific approach at the lowest possible level.[6]

Finally, what needs to be emphasized is that land use planning through UDCZs as well as through SLUPs, does not result from a specific approach nor is it placed within any institutional system of land use classification for the countryside. Also, it is not compatible with the existing agricultural land classification system such as the one devised by the National Statistical Service of Greece (NSSG), even at a coarse level (Table 6.1). On the contrary, UDCZs as well as SLUPs refer to land use categories or zones determined by the actual characteristics of the areas studied. Thus, surveys and plans that have been designed for the occasional, mainly by the Ministry of the Environment, Spatial Development and Public Works (MESDPW), are not comparable in that sense and are inadequate for the formulation of a unified land use policy (Economou, 1994).

Table 6.1: Land use classification by NSSG

Categories	Area (ha)
Crop areas and fallow land	3,945,200
Pasture lands (communal or municipal	3,00,380
Pasture lands (other)	2,254,640
Forests	2,951,090
Land covered with water	308,630
Built areas, roads etc.	489,310
Other areas	246,490
Total	13,195,740

Source: NSSG, 1981

Another framework for institutional protection of agricultural land was granted through Act 1650 'For the protection of the environment' issued in 1986. The Act included the option to define two basic categories of land use: 1) zones of nature and landscape protection (national parks, natural monuments, eco-development areas, etc.) and 2) zones for the development of industrial, mining, marble extraction, tourist or agricultural activities. Even though these spatial development regulations of Act 1650/86 do not define all the possible uses and procedures for an integrated spatial development regulation, they do reinforce the legal 'weapon' for the designation and management of sensitive areas.[7] Nevertheless, the administrative decrees concerning spatial development planning have not been issued yet, thus leaving the law inactive.[8]

In the remaining land that includes non-urban use and is not under the status of constitutional protection, any change of use is possible, with the consequent risk of uncontrolled urban expansion. Agricultural land, irrespective of its productive potential and in the absence of any complete protection scheme, is among these areas. The institutional framework for non-urban building, already formulated in the 1920s, allows for the construction of buildings of various uses[9] as long as the plots fulfil certain minimum conditions (minimum size 0.4 ha) and location (e.g., accessibility to main transport routes). However, an entire system of exceptions and deviations, defeats the main purpose of non-urban building regulations, which is to discourage building and to preserve the natural and rural character of the countryside (Getimis and Economou, 1992).

Among all the forms of land use covered by the legislation for non-urban building, houses used occasionally for recreation and tourism are the most common. As far as tourism is concerned in particular, one can detect an exceptionally favourable approach for tourist investments compared to other land uses,[10] that consequently affects the balance between them. For agricultural land such a regulation acts as an incentive towards further fragmentation of rural zones within which there is or is expected to be a demand for tourist activities. Often long before the realization of such investments, rural areas including highly productive agricultural land, are withdrawn from agricultural use and are either built on or put on the land market as *oicopeda*, i.e., building plots. Similar phenomena, though on a broader scale and more intensive in character, are encountered in recreation residential areas in an expanding radius around big urban centres, as well as in coastal areas close to inland settlements.

In conclusion, it could be said that the existing concessions for non-urban building lead to intensive land use similar to that in areas under urban planning schemes, yet without ensuring urban planning requirements, (i.e., master plan, technical infrastructure networks, public facilities). The diffusion of such types of building in vast areas of the countryside creates a so-called rural-urban fringe (Hervieu, 1987). As a result, in the entire country and especially where pressure for tourist development and recreational housing exists, landscapes that had been unaltered for thousands of years and are part of the national cultural heritage, have changed dramatically within the last few years.[11]

An assessment of spatial policies in Greece

Rapid urbanization during the 1950s and 1960s in Greece and significant changes in the agricultural economy during the next two decades, were all carried out without any systematic and restrictive framework regulating land uses, and without any spatial development planning. Greece has never had, until today, a plan specifying either the conditions for expansion of urban activities to agricultural areas or the requirements for changes in land uses. This is in contrast to most European countries, where spatial development and land use policy are essential parts of state policies.[12] Greek state policy has not been able to intervene effectively and orientate the competing forms of land use towards activities not derived from economic interests driven by an overvalued land property market.

Among these economic interests, small and fragmented holdings, which characterizes the structure of land proprietorship in Greece, have a decisive influence on the constitution of spatial development regulations. Indeed, many family farm holdings are made up of a number of small pieces of land (on average 4.3 ha in 6 parcels). Given the fragmentation of farm land, a certain number of plots is often located close to or within zones where expectations of urban, industrial or tourist development are high. Existing laws and regulations cannot guarantee the destiny of these overvalued lands. Under individual or collective pressure the authorities permit building, either legally or illegally, on former agricultural land. A significant social consensus has emerged, especially around the so-called 'unauthorized' building, transforming it into one of the cornerstones for the legitimization of the Greek political system. Consequently, the undifferentiated structure of land-use policies allows state agencies to take *ad hoc* decisions, such as the legitimization of unauthorized building, thus ensuring the reproduction of traditional clientelism, through the social consensus guaranteed by the 'sea' of Greek small landowners (Getimis and Economou, 1992).

Yet the absence of specific regulations for land use in the countryside, does not mean that there has been no policy at all. On the contrary, land-use policies in Greece since World War II have displayed considerable coherence and consistency in their priorities. Two characteristics of these policies, first substantial tolerance and even, in some cases, encouragement of illegal subdivision of rural land, especially of peri-urban farm fields, and second the frequent support for non-urban building, have been the main 'vehicle' for transferring agricultural land to various urban uses.

Another point that should be emphasized here and is directly related to the structure of land use in Greece, concerns the operation of the land market. It is not an exaggeration to say that the system of land prices constitutes the most important factor determining economic activities in an area, since the price of land derives from the most profitable activity. Competition between tourism and agricultural activities is an important case in point. Since the income from tourist investment is higher than from agricultural activities – which in land market terms means higher land revenue – the land owners are more often than not oriented towards the former. In practice this leads to expectation strategies usually involving the abandonment of farm land. Finally, the fact that broader economic policies work in a way which also enforces the ineffectiveness of the interventionist regulations, should not be forgotten. Land taxation and transfer tax in particular, are based on market mechanisms which are not directly compatible with environmental concerns. Tax is related to expectations arising from urban expansion and the consequent overvaluation of former agricultural land.

A number of recent attempts to reform spatial development planning have tried to overcome the problems and shortcomings of the general framework for the regulation of land use. These attempts started in 1976 with the Report for Land Use Policies issued by the Centre for Economic Studies and Planning, and later on with the new legal framework for spatial development planning and urban development (Act 360/76, Act 947/79, Act 1337/83). However, the report by the Centre for Economic Studies and Planning, did not include clear-cut options and never had a formal pattern to be put in place.

The same is true of other more recent attempts. Although they did address some reform issues, they came up against the structural problems of spatial development policies (doubtful proprietary status of land, claims on the same area by more than one landowner, etc.). They also met with the general reluctance of the state administration – central or regional – to draw up a specific spatial development framework.

Rural land-use regulations, incorporating a spatial development rationale, refer only to the SLUPs. Undoubtedly this attempt, though limited (the part of the country where rural land use is regulated is extremely small), precisely because it has taken place over the last few years, proves that the previous model of land-use administration of the countryside had started to cause serious problems that were noted at the level of the political system and at the decision-taking level.

Yet the limited number of SLUPs that reach the stage of enactment face another problem, namely the legal framework itself, that is the UDCZs. The procedure for enforcing UDCZs is extremely slow and bears a 'political cost' as the Greek politicians commonly say. This is due to the fact that the SLUPs/UDCZs almost always provide stricter regulations for non-urban than for urban building. Such proposals though, are met with powerful private interests (small land owners, building and tourist lobbies, etc.) and with interventionist practices of the Greek state. An additional aspect of the problem is the fate of SLUPs and/or of countryside UDCZs which have already been enacted. At this stage there are two risks: the loosening of existing regulations and the absence of effective control during their implementation.

I seems that the only meaningful tool enforcing the SLUPs, other than the UDCZs, is Act 1650/86. But even this Act does not constitute a legal framework directly oriented towards spatial development regulations. As has already been stated, the Act provides for the possibility of specifying areas with the main target being the conservation of nature and landscape. With the designation of these areas, the policy of nature conservation becomes more concrete because: a) it provides restrictions which cannot be imposed through the UDCZs (e.g., in areas of full nature conservation any kind of activity is forbidden); b) it allows for a set of financial mechanisms, such as subsidies, regular or one-off compensation payments, compulsory expropriation, etc., which limit the burden of the restrictions. Yet, the executive ministerial decisions that are necessary to activate these mechanisms, have never been issued. Thus, using the above zones as a tool for regulating land use is only theoretically an improvement over UDCZs.

Concluding this evaluation of land use within the framework of SLUPs/ UDCZs, it is worth mentioning the expected intensification of land use as it is regulated by the subdivision plot requirements. The average of subdivision ratio limits as well as other building regulations are slightly different from those for non-urban building.[13] It is obvious then that the difference between these reform attempts and the previous regulations is minimal, at least as far as the 'final outcome' is concerned. The only progress is that now more holistic and integrated research is being undertaken, which still remains useless to a large extent, since it is 'stripped' from all its initial targets before it reaches the final stage (Bieratos, Economou and Pyrgiotis, 1995).

Table 6.2: General land use and categories of countryside buildings

General land use in urban planning	Categories of countryside building
Pure housing	Agricultural buildings
General housing	
Urban centres	Industrial plants, offices, shops
Not disturbing industry – handicraft (middle and low disturbance)	Housing
Disturbing industry – handicraft (higher disturbance)	Schools, charitable institutions
Wholesale trade	Hospitals and clinics
Tourism/recreation	Tourist facilities
Open space in urban areas	Installations of Public Electricity Company
Public welfare institutions	

Source: Economou (1994).

The lack of a statutory land classification system is another important institutional issue. As an approximation to a classification of land use we can consider the list of categories provided by the official statistical service for the countryside, since they constitute the starting point for drawing up land-use plans (Table 6.2). Still, these forms of land use together with non-urban building activities, which basically account for the expansion of urban land use in the countryside, are determined by the logic of urban planning and building. Even the category of agricultural buildings, does not refer to agriculture but to the buildings themselves, which in this context are of secondary importance. As a result, the specific land-use plans describe *ad hoc* categories of land use which cannot in any way be reduced to a unified classification scheme.

Proposed alternative land use

Recently, there has been an attempt to develop a system of land-use classification. It relates to a joint proposal by the MESDPW and the Ministry of Agriculture for the classification of agricultural land (Table 6.3 and 6.4). As the tables show, the agricultural land of Greece – which today covers 3,933,600 ha – is classified according to use into three zones (I, II, III) and priorities (first, second and third priority). The main criterion is the degree of land-use protection. Therefore first-priority agricultural land includes land which should remain in that use in perpetuity. Second-priority agricultural land justifies less protection than the first, while the third zone includes unprotected agricultural land. The incorporation of 62 per cent of the Greek agricultural area into zones I and II has been based on objective criteria such as investment in infrastructure, irrigation prospects, the preservation of specific types of cultivation, etc.

Table 6.3: Agricultural land classification

Agricultural land zones	Area (ha)	%
First-priority agricultural land	1,600,000	41
Second-priority agricultural land	900,000	23
Third-priority agricultural land	1,433,600	37
Total	3,933,600	100

Yet, very often certain non-agricultural activities take place even on first-priority agricultural land, and therefore certain regulations and restrictions are necessary. All the same, this set of regulations has never been enacted, thus the problem

Table 6.4: Agricultural land classification by categories

Zones	Categories	Area (ha)	Location constraints
First priority	**A. Agricultural land already or to be irrigated**	**1,315,000**	Agricultural activities permitted according to existing legal framework conditioned by agreement of the Ministry of Agriculture.
	1. Within collective irrigation schemes	519,886	
	2. Outside collective irrigation schemes	663,114	
	3. Irrigation schemes under construction	132,000	
	B. Special agricultural land	**285,000**	
	1. Traditional and specialized cultivations	65,000	
	2. Others*	220,000	
Second priority	**C. Potentially irrigated agricultural land**	**486,500**	Non-agricultural activities generally forbidden, except activities specified by the Ministry of the Environment, Spatial Development and Public Works conditioned by agreement of the Ministry of Agriculture.
	D. Agricultural land under land reclamation schemes	**413,500**	
	1. Land reclamation schemes completed	216,600	
	2. Land reclamation schemes continued	55,950	
	3. Land reclamation schemes planned	140,950	
Third priority	**E. The rest of agricultural land****	**1,433,600**	All activities permitted conditioned by agreement of the Ministry of Agriculture.
Total		3,933,600	

Source: Ministry of Agriculture (1992)
*Cultivations forming natural entities, urban or peri-urban farming, areas of local importance, i.e., islands agricultural land.
**Marginal land, abandoned etc.

of protecting agricultural land remains and, as in other cases mentioned above, this does not seem to be accidental. It rather reflects the resistance that such attempts meet at all levels, whenever policies seek to impose building regulations and restrict intensification of land use.

Policies to conserve agricultural land, however, have another side. This concerns the anticipated harmonization of land-use changes with the directions of the new CAP which derive from 1992 reform. The opportunities available for such harmonization consist of a set of policies which, given the existing legal regime of land use, could reinforce the protection of agricultural land.

Take for example one of the most discussed issues in Brussels, concerning alternative agricultural production i.e., non-food (biomass). We tried to estimate the area available in Greece for such a policy, under given conditions. According to some yet unpublished data provided by the Greek Ministry of Agriculture, these areas do not amount to more than 268,000 ha in total, mainly of third-priority agricultural land (Table 6.5). Although these estimates are based on provisional and unofficial data, it is indicative that the area where alternative (non-food) production might be promoted does not exceed 7 per cent of total agricultural land in Greece, or 19 per cent of third-priority land.

Table 6.5: Estimated areas for alternative (non-food) production

Land use	Area (ha)
Market set-aside	18,000
Ecological set-aside	5,000
Pasture lands	15,000*
	80,000**
Abandoned land	150,000***
Total	268,000

*Released by the extensification measures (Reg. 2078/92);
**After the transformation of equal hectarage of marginal arable land to pasture land;
***Mountainous areas or located on Aegean islands of high tourist potential, such as Cycladas, Dodecanese or Crete, where the abandonment of farm land is rather irreversible under the existing regime of land-use policies

Source: Ministry of Agriculture (1995) unpublished data.

A second example is the implementation of Reg. 2078/92 itself. The schemes which are already under implementation (horizontal programme for biological agriculture, zonal programme for the reduction of nitrate pollution related to farming in the plains of Thessaly), as well as those that are to be implemented (horizontal programme for long term set-aside of agricultural land, which was approved by the Council in May 1996), cover an area of no more than 61,000 ha (1.55 per cent of total agricultural area).

Yet, besides the area that they potentially cover, what is of crucial importance with the implementation of all these new policies, is the message they carry for the need to reinforce the existing legal framework for the protection and conservation of agricultural land. What is more, the political position of those that support this idea is strengthened considerably.

Conclusions

CAP reform constitutes a challenge for Greek agriculture, in a period of liberalization of rural markets and consequent readjustments in the rural economy and rural policy. According to certain scenarios, within the next decade protectionism in European agriculture will be reduced radically. In its place compensation will be provided to the poorer farmers, decoupled from the level of production. Indeed, declining state support for the primary sector should be compensated by direct income support as well as by promotion of alternative use of farmland. Of crucial importance to this new orientation of rural development is the priority given to the protection of agricultural land from a conservationist's point of view. In Greece, institutional protection of agricultural land is extremely inefficient. The framework of political intervention in the management of land use is briefly described in this Chapter, while its weaknesses are related to the main historical and socio-political influences that brought them about. Among other drawbacks, the lack of an agricultural land classification system makes it difficult to develop a consistent, rational land-use policy that is easy to implement and control.

This institutional inefficiency is expected to impede the implementing of the new measures following CAP reform. As already mentioned, some of these measures support alternative management of farmland, while retaining its rural features. It is anticipated that the proposed payment schemes will strengthen the political will for institutional protection of agricultural land in a period of farming contraction. Or, in other words, overcome existing resistance to a stricter and clearer land-use policy. In order to weigh up the possible outcome of these contradictory tensions, we tried to estimate the area and type of land that competent state agencies (i.e., the Ministry of Agriculture) have planned to include within certain measures of the reformed CAP. As was shown, most agricultural land will not be covered by the new measures, while on the other hand, few opportunities are eventually offered within the existing legal framework for alternative use of these areas. Thus, despite its significance for the future of the Greek farming sector, the problem of agricultural land protection remains and is still subject to all those factors that prevented its solution in the past. What is more, because of prevailing trends in European agricultural policy, more pressure towards change in land use is expected, at the expense of its agricultural nature.

Notes

1. This Chapter draws from research done for the project on 'Large-scale biomass cultivation for energy in the EU: Evaluation of long-term impact on farm income, employment and the environment.' Project coordinators: S. Kyritsis and P. Soldatos, Agricultural University of Athens.
2. There are only two exceptions to this general rule; namely, archaeological sites and forest areas, for which there exists a specific legal framework established by the Constitution, that protects them and controls changes of their status. Yet, in exceptional cases incursions upon forests and forest areas are permitted, such as for land reclamation for agricultural purposes, road construction, building construction by cooperatives, tourist infrastructure works, mining activities, etc. In other cases changes are brought about violently (fires, illegal occupation) or through gradual adjustments in land use (from forest to pasture and/or agricultural land). The absence of a forest register makes it difficult to control forest land-use changes.
3. For example, agricultural land that is located within an irrigation network is protected from any change in use, only when it is included in a Master Plan (MP), or within the boundaries of an Urban Development Control Zone (UDCZs) or Special Land Use Plan (SLUPs). Yet when the designation of the MP or UDCZs, or SLUPs has not been put into force, then the agricultural land of the irrigation network remains unprotected.
4. Towards the end of the 1980s two successive sets of SLUPs, financed by EU, were assigned mainly to the Ministry of Environment, Spatial Development and Public Works. The first ones (1987–1990) came within the framework of Integrated Mediterranean Programmes and included 21 studies. The second ones, including 13 studies financed by the Programme ENVIRREG, have been drawn up from 1992 onwards, with the exclusive aim of determining land use for the development and conservation of coastal areas.
5. For the most part, the study areas are smaller than a prefecture, and usually include coastal zones of isolated regions or small islands. The areas for which regulations are finally proposed (intervention areas), are even smaller than the study areas, while areas for which the enactment of the regulations is expected – mostly through the UDCZs – are often smaller than intervention areas.
6. The only attempt at spatial development planning covering the whole country which has taken place so far, is the 'Proposals for Spatial Development of Prefectures' undertaken by the Ministry of Environment, Spatial Development and Public Works in 1984. These proposals however, have never been enacted. The parts that have been realized so far, even to a small extent, concern management of the urban development network, and the zones of environmental conservation. The latter were partly used later, in the elaboration of SLUPs, but as they have not been updated, the issues they refer to have become more and more unrealistic.
7. A special legal decision was issued in 1990 under Act 1650/86, also for the protection of the environment. This is the Common Ministerial Decision 69269/5387/1990 for

the 'Classification of works and activities into categories, content of environmental impact assessment, definition of the contents of special environmental studies and other relevant issues under Act 1650/86.' In particular, chapter E refers to protected areas and defines the content of specific environmental studies which are a prerequisite for the declaration of new areas under Act 1650/86 (art. 21, par. 1).

8. Another significant Act of the environmental legislation (Act 360/76 'For spatial development and the environment') remains inactive as well. Act 360/76 introduced a significant ordinance (art. 7) with very broad authorization as far as the development of certain activities or land use in specific zones/areas is concerned, according to the content of the respective spatial development plans. However, although this Act was approved by Parliament, the national or regional spatial development plans have not been issued yet, while the above-mentioned authorizations for the imposition of certain terms and restrictions have remained inactive since the Decrees under art. 7 have not been issued either.
9. These concern, apart from residential use, a series of other uses, mostly urban, such as industrial plants, office buildings, shops, agricultural buildings, hospitals, tourist facilities, public welfare institutions, etc.
10. It is worth noting that in a plot of 0.4 ha, which is the minimum allowed, it is possible to construct a tourist facility of 800 m^2., while for most of the other uses it is possible to build up to 200 m^2. What is more, another regulation, originating this time from the urban development legislation (Act 1337/83), includes a floor area ratio in reverse proportion to the size of the plot.
11. Recently though, with Act 2242/1994 there has been an attempt for a more systematic approach towards the urban planning of secondary residential areas within UDCZs, through the design and approval of a development plan for secondary residential areas. Beside the condition of not contradicting Acts protecting the environment, archaeological sites, forests, etc., urban planning should be in accordance with existing regulations of spatial development at prefectural level, the need for the preservation of highly productive agricultural land and the 'balanced' use of the seashore by taking into consideration their morphology and aesthetics.
12. Spatial regulations in these countries mainly relate to such issues as the systematic control of land use as well as land-use changes, enactment of spatial development plans in relation to development programmes, incorporation of investment plans for infrastructure into a framework of long-term spatial development targets, intervention of land banks, control of any extension of urban use in agricultural areas, etc. Although today many of these policies – formulated in the 1950s – have lost their general and 'integrated' character of state control and have become more selective, fragmentary and open to market processes, still the framework of spatial regulation in these countries is much more powerful and effective than in Greece.
13. As an indication, the subdivision limit for housing in most of the enacted UDCZs is just over 0.5 ha, for agricultural land between 0.6 and 0.9 ha (the higher rate is for high productivity agricultural land), for industry 0.4–0.6 ha. etc.

References

Baldock, D. et al. (1993), *Nature Conservation and the New Directions in the EC Common Agricultural Policy*, Institute for European Environmental Policy: London.

Bieratos, Ç., Economou, D. and Pyrgiotis, Y. (1995), *Classification and Controlling Mechanisms of Land Uses in the Countryside. Comparative Study of Other Countries' Experience. Proposal for Greece*, Research programme, University of Thessalia, Department of Spatial and Regional Development.

Economou, D. (1994), 'System of Spatial Development Planning and Special Land Use Plans', *Environment and Law*, No. 1, pp. 41–86.

Getimis, P. and Economou, D. (1992), 'New Forms of Geographical Inequalities and Spatial Development Policies in Greece,' *Ôopos*, No. 4, pp. 3–44.

Hervieu, B. (1987), 'From the Disappearance of Peasants to the Renovation of the French Rural Social Territory: French Sociology under Dispute,' *Review of Rural Studies*, No. 2.

Ministry of Agriculture (1983), *Guidance for the Application of Act 1337/83*, Department of Environmental Protection, Ministry of Agriculture: Athens.

Ministry of Agriculture (1992), *Legal Protection of Agricultural Land. Proposal to the Ministry of Environment Spatial Development and Public Works*, Department of Spatial Development and Environmental Protection, Ministry of Agriculture, Athens.

Poux, X., Baldock, D. and Mitchell, K. (1995), *Preparatory Study for the Consultative Forum on Environment. Setting Policy Scenarios for a Sustainable Rural Development*, Brussels.

7 'We can all use calculators now': productionism, sustainability, and the professional formation of farming in Co. Meath, Ireland

Hilary Tovey

Evaluating the traditional in 'traditional farm culture'

There is an argument quite frequently encountered, in rural sociology and in the writings of other groups concerned with the rural environment, to the effect that the experience of sustained pressure towards a productionist orientation to farming, coming primarily from European and national policies for agricultural development from the post-war years until the mid- or late 1980s, led to the suppression of an older cultural orientation towards the land and towards nature. This earlier orientation found expression in a discourse of 'good husbandry' or 'landstewardship,' which seemed to recognize limits to the extent to which natural resources could be exploited in the process of food production without needing to be replenished, and to favour forms of farming which worked within these limits – in effect, which were 'sustainable.' Productionist orientations on the other hand, it is argued, see no natural limits to production (limits are set by markets but these can be postponed for a long time by appropriate forms of state intervention) as long as farmers are prepared to intensify and commodify their inputs, which also meant that they had to accept packages of production practices which were to a large extent created outside of and dependent on bodies of knowledge inaccessible to local farming communities. A further implication often drawn from this is that if contemporary agriculture is to become more ecologically aware and more 'sustainable' in its practices, this must involve somehow a return to and revitalization of the earlier traditional culture and practices; environmentally sustainable farming is not compatible with the cultural orientations associated with productionism.

Thus, for example, Miller has referred to 'the current orthodoxy in rural research (perhaps best synthesized by Symes in *Sociologia Ruralis* 1992) that the trend of postwar productionism has entailed an eclipse of more traditional indigenous cultures approximating to a kind of "custodial" approach to the land and landscape . . .' (1995a, p. 2), and elsewhere, to the argument that 'there has been a change in culture as well as in economy at the local level . . . the productionist revolution has brought about the abrupt decline of a rural world of practices and values, loosely designated "traditional," and induced instead its antithesis, a system of agribusiness driven exclusively by profit-maximization' (1995b, p. 101) which he understands as widely shared in the contemporary literature with only one or two 'lone voices to the contrary' (ibid).

Outside rural sociology we find similar ideas. Thus, a study of Irish farmers' attitudes to the countryside (Reidy, 1992) operationalizes environmental awareness among farmers in terms of three measures: agreement with the statement that farmers have a role as guardian or protector of the countryside; willingness to become involved in environmental management agreements; and a favourable assessment of the skill requirements of 'traditional farming practices.' Throughout his discussion Reidy treats 'traditional farming practices' as so obviously identical to 'farming with environmental awareness' (1992, p. 133) that the relationship does not need to be elaborated. His data are presented to show that the farmers most opposed to traditional practices are those with large farm operations and particularly farms specialized in intensive dairying, tillage, pig or poultry production – that is, those who have been most exposed to productionist perspectives on agriculture, as predicted by the argument outlined above.

One thing which is happening here is an interesting re-evaluation of 'traditional' rural culture. Seen during the productionist era as an obstacle to agricultural modernization and commercialization, and more broadly in societal terms as the prime representative of that 'backwardness' which it is the task of 'the civilizing process' to destroy (de Haan, 1995), traditional culture has now become the focus of a variety of attempts to reconstruct contemporary society in order to deal with the problems of social solidarity or of ecological degradation to which modernity gives rise. In rural sociology the revived interest in rural culture, as de Haan points out, is associated with a range of other developments – the rebirth of social movements around regional or local identities, thought to have been finally vanquished by modernization; theoretical and empirical work on 'local knowledges' and their capacity to resist 'scientization' (Van der Ploeg, 1992; Murdoch, 1994); a growing interest in theories of post-modernism and their implications for the interrelations of the local and the global in constructing individual identities. What is perhaps more striking still is the re-evaluation of 'traditionalism' among groups in society who would previously have been most committed to modernism. Reidy's (1992) paper is interesting not only for the particularly clear way in which it expresses the assumptions about productionism, traditional agriculture and sustainability with which we started out, but also because as an employee of Teagasc/the Irish Agricultural Institute, he works within the institution most directly charged by the state for many years with the task of eradicating all but economic rationality in Irish farmers' considerations of how to manage their farms.

The re-evaluation of 'traditional culture,' then, connects with some of the more dynamic and interesting developments in contemporary rural sociology. But it is worth considering why the concept lost value in the first place. The problems connected with the term 'traditional,' pointed out for example by Frank (1969) in the 1960s, still remain, particularly if it is intended to suggest that cultures can be 'traditional' in the sense of being outside historical time, transmitted more or less unchanging over many generations and relatively immune to the effects of societal conflict, variation or transformation. This way of understanding traditional rural culture, for example, dominates much of the work conducted within the French rural enthnological school, as in Zonabend's account of how a French village remembers its past: 'The time of the community shuns history . . . Between the account of the village's origins and the middle of this century nothing has caught the attention of the community, which considers itself to be outside history, outside chronological time, but existing in space . . .' (1984, p. 3). Anthropological work on rural Ireland also, particularly Arensberg and Kimball's *Family and community in Ireland,* often represented western peasant farmers earlier in this century as embodying a culture which existed from time immemorial, although objectively the conditions which allowed the particular forms of farming and family formation, marriage and inheritance customs and labour relations with neighbours to develop could scarcely have existed for more than two generations in the locality studied. Such work has been passionately disputed by Irish sociologists as a-historic and excessively consensual (see Tovey, 1992 for details).

A second problem is that a rural culture described as 'traditional' is too easily treated as homogeneous and unchanging across space, as well as across historical time. To take an example again from Irish rural sociology, the findings of studies of peasant farm cultures in the West of Ireland have often been assumed to be transferrable to other parts of the country, on the grounds that they portrayed a universal traditional rural past. This chapter is, among other things, a challenge to that assumption. Much recent work in sociology and social geography suggests that in fact we should expect farming cultures to vary substantially from one locality to another. Van der Ploeg, for example, emphasizes how agriculture has produced 'a highly diverse mosaic' (1992, p. 19) across space as a result of the localized conditions which give rise to locally specific farm labour processes and mechanisms for their reproduction. Thus in the past, he argues, 'Every location acquired, maintained and enlarged . . . its own cultural repertoire, its own norms and criteria that together established the local notion of "good farming"' (1992, p. 20). Similarly Johnston argues that 'the local milieu' plays a fundamental role in structuring how people respond to problems, and that as culture is formed out of attempts to manage life problems, thus 'cultures develop in places and are passed on in places' (1991, p. 50). If we accept that farming localities in the pre-productionist era were characterized by the presence of a strong local culture, then we should at least not apply the term 'traditional' to them, recognizing that they are precisely not outside historic time but a response to quite specific sets of historical social structures and historically conditioned environmental processes which are part of the specific experience of people living in a particular social and physical space.

However the case which this chapter examines – that of large, modern farmers living in the east region of Ireland, in County Meath – takes the issue of local cultural variability even further. It is argued here that these farmers do not appear to have any significant sense of a past local culture of farming at all, either as something with which they claim continuity today or as something which they have rejected. The argument is not that there was no earlier culture of farming to be found in Meath – although the social structure of the area seems likely to have worked against the emergence of a single, shared, local culture. What is argued, rather, is that the example of Meath indicates that there are at least some cases when, to understand the contemporary practices and attitudes of farmers, it is not in fact useful to appeal to some hypothesized past culture whether as a source for or a contrast with the way they understand farming today. The discussion sets out to show that we can understand the culture of Meath farming today better if we take Meath farmers as oriented not to the past but to the future: it is their visions of the future of farming in the area which largely shape their interpretations of what needs to be done now, and how it should be done.

Discussion of the Meath case suggests two further points. First, that the absence of a sense of a shared cultural past among Meath farmers has itself been one important local condition for the emergence of a strong contemporary collective identity, which has been formed around the professional role of farmer. (Other significant local conditions include the decline in heterogeneity and social divisions among the local population, the size and the commercial history of holdings in the area, and the early political mobilization and participation of farmers in this part of Ireland). Second, this professional identity is strongly centred on farming practices and understandings about good farming which would generally be considered to be 'productionist.' But the case study reveals that this does not preclude the contemporary farming culture in Meath from containing attitudes toward domestic animals and wildlife, towards landscape and the local ecology, which are compassionate, appreciative or aesthetic. And it appears also that adjusting to the changing demands of policy makers and the general public – including demands for new attitudes towards animals and the environment – is understood within Meath farming culture as itself a core characteristic of the professional role of farmer. These points suggest that if we are looking for ways of encouraging contemporary farming towards greater sustainability, these need not take the form of trying to make them return to 'traditional farming practices' or 'traditional culture': the elements necessary for a shift to sustainable farming may already be present in their contemporary, even productionist, cultures.

In the paper cited earlier, Miller criticizes rural sociology in Britain for relying too much on a discourse about farming and the rural which has been constructed by, and predominantly reflects, urban world views and interests. He suggests that the attitudes to nature and the sense of place of rural people themselves have largely been ignored by researchers, and that what material is available may tell us 'more about establishment constructions of rural life than it does about the local values and ethos of active farmers' (1995b, p. 102). This is particularly crucial in relation to 'the purported transformation' in farming culture 'from "traditional" to "agricapitalist",' where, Miller argues, 'We need to go beyond these schematic generalisations of cultural change and examine . . . the empirical detail of the assumed transformation at the local level or "locality"' (1995b, pp.101–2). Rural sociology in other European countries, including Ireland, may be less empty of empirical rural detail than Miller suggests for Britain; nevertheless this chapter can be read as a step in the direction which he recommends.

Studying Meath farming culture

The farmers described here – the relatively large, well-off producers of Co. Meath on the east side of Ireland, just north of the capital city of Dublin, who are interested in and able to afford science- and technology-based innovations in production and who are closely attuned to changing market conditions – occupy a rather distinctive historical and (if to a declining extent) contemporary position within agrarian Ireland. But they have been surprisingly neglected by Irish rural sociologists whose research has been almost totally confined to farming in the more western and south-western regions of the country. Irish rural sociology has been dominated by an understanding of contemporary Irish farming as something which emerged out of an earlier 'peasant' economy and society (Hannan, 1979; Curtin, 1986). Research and debate has focussed on the process of social change which is taken to be the historical context within which today's farmers must be understood, and which is understood as the collapse, from the 1960s on, of 'peasant' farming and of the restrictions that peasant communal norms placed on entrepreneurial forms of behaviour. Both those 'family enterprises' which today are modernized and productive, and those marginalized landholders who have been unable to develop into full-time modern farmers, have thus been theoretically located in the same 'post-peasant' agrarian social structure. However the thesis succeeds as an universal account of development in Irish farming either by ignoring the rather different developmental history of farming in the East or by assuming, in the absence of detailed research, that the commercial family farming found here is also a 'post-peasant' form but that the peasant society out of which it emerged was destroyed at a much earlier stage.

When the opportunity arose, in 1993–4, to carry out funded research on a group of Irish farmers, we chose to focus on Meath farmers partly because of their neglect by other Irish researchers, and partly because we wanted to study productionist orientations towards farming and farm development. Meath farmers were likely to have been exposed to these earlier and more intensively than other Irish farmers and were more likely to have adopted productionist practices on their farms. We identified an area in the mid-west of the county where introductions to the type of farmers we wanted to study (predominantly large cattle finishers, but also some dairying, tillage and mixed sheep enterprises) could be organized, through the help of the local office of the state-organized agricultural advisory service, and the research assistant on the project moved to live in the area for a period of some 12 months. The research process thus combined ethnographic fieldwork in the area (through repeated encounters with a small number of farm families with whom friendship relations developed, and visits to local branches of various formal organizations such as the Irish Farmers' Association and the Irish Countrywomen's Association) with over forty tape-recorded interviews with farm households, most at least two hours in length. The interviews usually began, often in the farm kitchen, with 'the farmer' (almost inevitably identified with the male 'head of the household'), then expanded to include farmers' wives and also children, if they were present and prepared to join in. In a number of cases a further interview was conducted outside the house while walking around the farm, visiting animals in different fields or being shown what the farmer considered to be particularly interesting features of his land, landscape or farming practices. Between interviews and ethnographic fieldnotes, a large body of material was generated. The discussion here draws on only a selection from that material, seeking out what appears to be the most 'relevant' for our purposes in the chapter.

The interviews were conducted in as unstructured a manner as possible. We had prepared a list of issues on which we hoped to get our interviewees to talk – for example, their views on contemporary farming and the differences they see between this and farming in the past, their own career history as a farmer, the sorts of concerns which guide their decision-making processes on the farm, their involvement in farming politics, their attitudes towards new claims being made on the countryside by non-farming interests – but we tried to allow these to arise naturally in the conversation rather than introduce them directly. Our interest was primarily in the present, in identifying contemporary productionist traces in the talk of the farmers, but we took for granted that we would be able to use conversation about the past as a non-threatening way of moving into questions about present conditions and practices. So it came as quite a surprise when we discovered, on reviewing the material later, that it yielded very little that we could identify as a collective sense of a shared Meath farming past. This led to a reconsideration of the role which is usually accorded to the past in shaping people's present understandings, and to an attempt to develop an alternative way to approaching the Meath case.

Farmers, and country people generally, are conventionally expected to have a strong sense of the past, because they are thought of as typically a relatively immobile group, socially and geographically, who have therefore been able to construct and share a collective interpretation of their past history, and perhaps also because of an assumption common among urban groups that people who live 'close to nature' are also 'close to their roots' as people. But there are also theoretical arguments for the view that people in general invoke a shared account of the past in structuring their talk about the present. Connerton, for example, investigating 'how societies remember,' states that 'our experience of the present very largely depends on our knowledge of the past' (1989, p. 2) and that it therefore differs according to the different pasts with which one connects it. Hence control over a society's memory and knowledge can yield great political power. Later he adds that 'Images of the past commonly legitimate a present social order. It is an implicit rule that participants in any social order must presuppose a shared memory. To the extent that their memories of a society's past diverge, to that extent its members can share neither experience nor assumptions' (1989, p. 3). Johnston rather similarly argues that the distinctiveness and boundariedness of contemporary local cultures is rooted in the 'collective memory' (1991, p. 50), which is the carrier of previously worked out local solutions to local problems, the 'store of knowledge' on which local groups draw as they face the problems of survival.

If the Meath farmers to whom we talked did not reveal communal remembering, then, these theoretical approaches suggest that it must be because we failed to look in the right places for it in the research process. Connerton argues that there are two central locations in which 'communal memory' or 'tradition' should be sought: in 'commemorative ceremonies' (the focus of his own discussion) and in 'informally told narrative histories' which are also an important constituent of social memory. Our field research did not include local ceremonies of commemoration and our impression is that these played little part in the collective life of Meath farmers, although of course they can be extremely significant within individual families. But the absence also of informal narratives about a communal past seemed to indicate that the problem was not simply one of inadequate or incomplete research, but of the theoretical construction of behaviour underpinning approaches like those of Connerton and Johnston.

While it is true that, as Connerton claims, political legitimation feeds on shared interpretations of the past, it can also derive much of its force from shared expectations about the future, as the mobilization of different social classes around the goal of economic growth in any 'developing society' indicates. Connerton's approach, moreover, imports unnecessary presuppositions about social consensus in society. He addresses the problem societies have of 'communicating incommensurable memories' (1989, p. 3) across generations to secure continuity of participation, but does not address the problem within generations. Yet it is, arguably, a normal feature of stratified societies that people from different social groups have incommensurable memories, yet can continue in some sense to 'participate in the social order.'

Schutz has suggested (1971) that the phenomenology of social action includes two radically different ways of imputing meaning to what individuals do: 'because' motivations, which are the typifying interpretations of action made by external observers and which treat action as structured by reference to the individual's past biography, experience and social location; and 'in order to' motivations, which refer to how the actor him or herself experiences action, as a striving towards realization of some future purpose or goal, or an attempt to meet an obligation, for example. While Schutz treats these as 'external' versus 'internal' accounts of action, we might instead suggest that they reflect cultural or historical or local differences in the way people locate themselves in relation to their external circumstances. In some circumstances local culture encourages a sense of oneself as a creature of biography, history and social structure, in others, a sense of oneself as their creator. The latter is of course a central element in the culture of modernity itself (Giddens, 1991). This suggests that in some cases we may make better sense of a farming culture by treating it as a localized form of modernity, than as a local variant of tradition.

Remembering the past: family history versus communal memory

Meath farmers did of course talk about their past, though not about a collective past. For example, people who had moved into the county to farm tended to emphasize that fact of their biography. One man who was born in Co. Monaghan, and came to Meath as a young farmer starting out, said:

> Well, I'm more than half my years here, but no matter what county you go in, they always say to you, you know, well, the county you were born in, you never get away from that.

Farmers like these tended to know little about the history of the house they lived in or the lands they farmed. A fairly typical in-migrant statement is the following:

> I wouldn't have a lot of the history of the place . . . I think the family of the original house – it was away down the fields in a little corner, now who lived in it I don't know, you'd have to go down to the parish register or something to find out . . .

Farmers who were born in the locality were also unlikely to offer much detail on its past history. Their comments generally went back only one generation, covering for example how they remembered the area from their own school days and their contacts in school with the children of other locals. But for both groups of farmers, immigrant and local, family history rather than the history of the locality or of Meath farmers as a collective group was what provided the structure of memory and past time. Farmers and their wives talked of the past primarily in terms like 'when I first took over from my father,' 'when we were first married,' 'when we first moved to this farm,' or 'how my father used to do things, when I was still at school.'

A striking feature of such narratives is the lack of historical depth even in their talk of their family history. It was extremely rare for any respondent to offer comments on life in his grandfather's time. Even fathers' lives were not discussed in great detail but invoked mainly as a context for explaining some aspect or other of the speaker's own way of doing things, or of his present or recent situation. One cattle farmer, for example, who clearly took great pride in his entrepreneurial skills which he believed had allowed him to build up a much bigger holding than his father had, told how as a youth of 17 or 18 he wanted to buy a tractor (they had four horses on his father's farm but no tractor) and needed the loan of £300. He put it to his father that if he got some help getting the loan he could repay it very quickly by hiring himself and the tractor out to work for other farmers, but 'It was like a red rag to a bull . . . I asked him to secure the loan for me but no way, a neighbour had to give it to me.' Others used their father's experiences to make rueful comments on how they themselves had failed to avoid the same sorts of troubles, as in the following comment from a specialized grain producer:

> Confidentially, I have always had trouble with banks and high interest rates, historically since the late 1970s we have been paying up to 30 per cent interest rates on and off over the years, and some of those high interest years coincided with bad weather harvests and it's amazing how often that happened . . . I remember my father saying to me, you know, he'd had a bellyfull of banks by the time he was retired. Myself and my brother were saying the other day, once we clear out our problems now we'll never borrow money again.

But such comments rarely suggested that the speaker was trying to make general points about farming in Meath, past and present; nor was it possible to find in them any consistent and collectively held thesis about that farming and how it might have developed or changed over recent decades.

On a few occasions interviewees did talk about the past in words which hinted at elements of a larger than family narrative. An example is found in the following extract from an immigrant – or to use the common Irish phrase, 'blow-in' – farmer, who started with a very small holding, around forty cattle, and now has a herd of 250:

> It wasn't very advanced at all, in this part at least, until the last number of years. So much has changed in Co. Meath. I don't know if this would interest you but I saw some *very* big farmers when I landed up here, you know, here I am in this place, trying to develop it, and I could name, oh, five or six of those big farmers, from 600 acres [c. 1500 hectares] or upwards, and they haven't survived, and the little lads, that had no money, came up, you know, little 30 acre [70 hectare] farms and eh honest to god they've outdone them. I've seen that . . .

This picture of Meath farming as a meritocracy where the 'little lad' who had the right skills and attitudes could end up by outdoing the longer established and 'too comfortable' large landowners of the area could be interpreted as a reference to a pre-productionist style of farming which was bypassed and left behind during the productionist years. But two points can be made about it: it does not occur frequently in the interviews, and when it does it seems to be related to a specific class-based or even nationalist-based narration about social change. This is discussed further in the next section, which suggests that the agrarian society of Co. Meath in the 19th and early 20th century has been widely regarded in Irish society as a particularly clear example of a colonial social system, deeply stratified along ethnic and socio-religious lines. In nationalist accounts of the past, political independence is seen as having brought about the destruction of this almost 'caste' system, and opened up access to resources such as land to any Irish person who had the talents to take advantage of the opportunity.

Insofar as the farmers we interviewed were prepared to construct accounts of 'the past' at all, then, in talking to us, they either talked about individuated pasts, as seen through family histories and in particular the succession of father to son within agriculture, or else about a collective past which could be the past of the 'Irish nation' in general, understood as a nation of small-scale farmers who overthrew caste and class barriers to success. In the next section various ways are explored in which that might be explained.

Agrarian society in Meath – from differentiation to homogenization

One possible explanation for the absence of a shared view of history among the Meath farmers needs to be discussed further immediately, because of its evident plausibility. It is that they have no subjective view of themselves as a historic collectivity because objectively they did not constitute a collectivity in the past.

I suggested earlier that there are problems in understanding Meath farmers as part of a 'post-peasant' society. The historical forms of farming in Meath, in terms of relations to land, to labour, and to the product produced, appear very different from those in other, particularly more western and north-western, parts of Ireland. To understand these forms we have to remember that until the 1920s Ireland was a colony of Britain, and that it followed the classic colonial dependency pattern, in terms of its economic relationships, long after gaining political independence. Its dominant markets were located in Britain, to which it mainly exported relatively unprocessed agricultural goods, and from which it imported industrially produced items. In the early 19th century much of the land in Co. Meath was already used as the basis for a highly commercialized export-oriented form of beef production, better called 'grazier' or 'rancher' farming than 'peasant' production. Land was owned by a small number of very wealthy and often absentee landlords, from whom it could be rented on short-term leases and used to store large numbers of cattle which had been bought from smaller tenant farmers in the West and midlands. The system suited the landlords who were interested in reducing the number of sitting tenants on their land, and suited the graziers who could take in or let go land depending on conditions on the English market for beef, and who needed to invest little in improving the land they rented as it was not primarily used to rear stock but rather to hold them while waiting for the most favourable market conditions (Jones, 1983).

When the change in land tenure from tenant to owner operator was introduced at the end of the 19th century, larger and richer tenants in the eastern parts of the country were the first to take advantage of it, even though it meant that a degree of flexibility in the production system would disappear. The advent of 'peasant proprietorship' has often been seen as allowing Irish farmers to retreat, in the 1920s and 1930s, from commercial into more subsistence-oriented types of production; but in the East of the country land markets remained active and the new owners of the land seem from an early stage to have taken a quite economically rational attitude to their holdings, selling off one farm to buy another if it seemed likely to realize more profit, or indeed buying into farm land as an investment at certain times and shedding it at other times when other more profitable opportunities became available. The attachment to 'keeping the name on the land,' or intense feelings about possession of specific piece of land, which marked other farming areas until the most recent generation appear not to be marked features of Meath farming. Whereas in the West of Ireland local social stability across the generations since the mid-19th century was mainly a product of 'peasant' farm and family systems, in the East it is more likely to have been the farm labourer class than the farmers who provided this stability.

Meath has long been regarded as a very rich farming area. Particularly in more southern parts of the county, grass grows lavishly on deep soils of heavy clay which also support a profusion of trees and hedges. The landscape is not dramatic or spectacular – much of it is flat, and the hills that are there are generally rounded and gentle. It is a thoroughly 'civilized' landscape, clearly marked by the centuries of human intervention and cultivation that have gone on within it. According to one writer, 'In the popular culture of Ireland and especially among the nation's farmers, Meath is rightly considered to be the county with the greatest number of cattle, the most fertile land, the largest farms, and the richest farmers' (Wilson, 1988, p. 680). Popular urban culture, also, identifies Meath as the county of the 'strong farmer,' wealthy, well connected, conservative and politically powerful.

In some ways this popular characterization of Meath farmers is misleading. In the early 1980s, the county had the same proportion (56 per cent) of farms which were less than 75 hectares in size as the whole country did. But it is true that a striking characteristic of Meath farming today is the presence of a considerable number of what are by Irish standards very large holdings (500 hectares and above), and that these have tended to increase more rapidly here than in other equally wealthy farming areas in the midlands or South. Their increase is due not only to the incorporation of small holdings into larger farms, but also to the expansion of small holders and the break-up of the very largest holdings (1,000 hectares or more). In terms of scale of farming, in fact, we could argue that Meath farming has moved from being highly differentiated to becoming rather more homogeneous over the past half century or so. More farmers in the county nowadays actually belong to that social class – the agrarian bourgeoisie – to which outsiders are inclined to consign all Meath farmers.

Culturally, they have also become more homogeneous in recent decades. What are today classed together as 'large Meath farmers' came in the past from at least three distinct groups, differing not only in their class origins but also their religious and ethnic group membership, and their political orientation. Up to the 1960s or so it was still possible to identify in Meath a last remnant of the 'county families' of the pre-Independence period, descendants of the Protestant Ascendancy whose lifestyles and attitudes probably had more in common with their class equals in Britain than with other groups in rural Ireland. By the 1960s these had lost most of their power in the countryside although their way of life could still command some deference as a role model from ambitious members of the urban bourgeoisie. None of these were included in our sample though we did encounter a small number of large farmers who could be called 'middle-class Protestants,' descendants for example of agents employed by the landed gentry to manage their estates, who themselves moved into farming in the late 19th or early 20th century. In the 1940s and 50s these would have lived in a world very separate from that of their Roman Catholic neighbours, not because they defined themselves as 'British' rather than 'Irish' but simply because they belonged to a separate parish and social network with different leisure and socializing habits. Protestants would rarely have drunk in local pubs or attended Gaelic Athletic Association fixtures, for example, nor would they have taken the same part in local government as Catholic large farmers did at that time.

By the 1990s such social separation has largely disappeared, and in any case the great majority of large Meath farmers today are Roman Catholics. But even these have not until recently been a homogeneous status group. While some of them (for example, descendants of some of the large tenant farmers of the last century) have a long history of residence in the area and of participation in, indeed domination of, local politics as county councillors and as leaders of local organizations, others are people whose land ownership in the county is of quite recent origin. Our interviews identified a number of farmers who were either themselves immigrants from the midlands or West of Ireland or the children of such immigrants; often brought into the area by grants of land from the Land Commission, they went on to build up extensive holdings, often by buying out the farms of other grantees who failed to make a success of their new situation. However as Meath farmers have withdrawn from leadership of local political life, since the 1970s, and become more involved in the politics of farming, nationally and internationally (Wilson, 1985), such differences are of declining importance.

Thus the historical experience of Meath farmers has been that of a movement from diversity towards homogeneity, and we could argue therefore that any collective consciousness of themselves as farmers which may exist can be of only recent formation. Yet the objective fact of earlier diversity does not necessarily preclude groups from constructing a collective past to legitimate their present, (Connerton, 1989) or from developing what Anderson (1983) has called an 'imagined community,' which includes myths of origins and other references to a symbolically meaningful past. As mentioned in the previous section, it seems that in the Meath case any imagination of past communality that may exist refers back to a constructed history which is more national than local. It may be that rural sociology is inclined to overestimate the extent to which farmers are captured by the local and to underestimate the extent to which, under certain circumstances, they understand themselves as part of 'society as a whole' and their history as the history of national society. This itself may be one significant way in which local rural cultures vary, across space and time.

There is one respect in which Meath farming in the past fifty years or so has become more diversified rather than more homogeneous: this is in relation to the commodity produced and hence the type of farming operation in place on the farm. Particularly since entry into the then EEC in 1972, beef production has been losing its dominant position in Meath. Two new trends became established over the period. Some farmers, generally the largest landholders, turned towards specialization in tillage crops (the main ones being winter wheat, oil seed rape, and potatoes). A larger number, if still a minority, switched into intensive dairying. Some farmers who had previously fattened beef stores for sale moved instead into rearing sucklings, and many started to keep large flocks of sheep, usually alongside their other, main enterprise. Land use in general in the county has become more intensive, with a steady decline in the amount of land left fallow, under woodlands or scrubby growth. The growth in tillage has not produced as much of a contraction in land under pasture and rough grazing as might have been expected; instead, as land values rose steadily, it became economic to bring previously 'marginal' land into production. Contemporary set-aside policies are starting to reverse this trend but as we see later, with little support from farmers.

The fact that farm operations were more homogeneous in the past, and have become increasingly diversified, is interesting given that despite the apparent absence of a collective sense of the past, a collective discourse about farming today does appear to exist among Meath farmers. It is one in which differences between different commodity producers – between dairy farmers and cattle finishers, for example, or between cattlemen and tillage farmers – are often commented on. Our data suggest the presence in Meath of the sort of 'social mapping' of different local farming styles and degrees of farm commoditization which is discussed by Van der Ploeg (1992, p. 29) as a basis for establishing the continuing significance of local knowledges in the face of scientization. In our case however it is less detailed and is attached to different lines of commodity production rather than to different styles within a given commodity line. Our interviewees were clearly aware of the different interpretations attached to each. Dairy farmers, for example, often described their form of farming as the most demanding in terms of the skills and quality standards demanded; on the other hand, tillage and beef producers were quite likely to refer to this conception which dairy farmers had of themselves in amused or exasperated tones. One cattle producer recounted how a salesman had come into his yard a year or two before to try to interest him in a new product:

> He says to me, "You're not in dairying?" I said no, dry stock and suckling – "Oh, you're not busy so" . . . There is an association between dairying and being busy and between fattening cattle and not being busy, but I reckon I'm just as busy and have to put in as many hours as the best dairyman.

The mutual knowledge of each other's stereotypes which such comments indicate suggests that the farmers we studied share a common conception of their present. It is argued in the next section, however, that this conception is one which is oriented towards the future, not the past. It is thus both an outcome of and a condition for the absence of even an 'imagined' local communal history among Meath farmers.

The professional formation of Meath farmers

Since the 1950s, but especially since the early 70s, which would be within the farming lifetime of most of our subjects, the large producers of Co. Meath have become increasingly organized in terms of their profession as farmer. Moreover, since EEC Accession in 1973 they have increasingly understood themselves as part of a profession which is established not just nationally but also internationally, and in this context local differences in religion, class of origin or length of tenure within the county, and so on, have lost any significance as barriers to collectivization. Differences in commodity line have also become more pronounced; but they have not constrained the emerging collective identity of 'farmer' since they are themselves the stuff out of which professional identity is created.

Two agencies can be picked out as being central to the formation of a professional identity. One is the state; it has always been involved in managing Irish agriculture, but from the 1950s onward first the Irish state and later the EC became explicitly committed to bringing about certain sorts of structural change through the use of commodity supports and farm development schemes. From the 1960s to the 1980s state agents in Ireland tried to stimulate adoption of a particular style of farming, based on intense use of commodified inputs and technological know-how to secure a high volume of output, in order to solve what was taken to be the central problem in Irish farming, low incomes. Accompanying this were measures to reduce the number of farmers (for example via the Farm Modernization Scheme) to only those whom state agents considered capable of generating a viable income from their farm, and some attempt to provide alternative off-farm jobs for those who were not thought capable of doing so.

The second is the Irish Farmers' Association (IFA). The IFA, the main (though not the only) representative organization of Irish farmers, was established in the early 1950s, but it was not until the 1960s that it began to exercise a significant influence over the way farmers thought about themselves collectively, when the balance of power over state development policy first started to shift from 'rural' towards 'urban' (more accurately from agricultural to industrial) interests in Ireland. State determination to increase industrial employment by subsidizing foreign companies which located branch plants in Ireland left farmers fearing that the amount of state funding allocated to agricultural development would be progressively reduced. After a series of mass protests and demonstrations in the 1960s, which the IFA organized ostensibly to highlight the plight of small-scale cattle producers in the West of the country who were suffering from a catastrophic collapse in calf prices, by the early 1970s the IFA had established its right to be treated as the main representative voice of Irish farmers, admitted to corporatist relations with the state, and eventually included in consultation on national policy formation along with the main urban-industrial social partners.

Meath farmers, as producers of cattle for export and hence an important source of foreign earnings for the state, had some success in the early years of the foundation of the state in influencing agricultural policy to favour the interests of large landholders above those of small-scale farmers. For example the costs associated with ownership of land were reduced virtually to nothing, and access to development capital was provided at a significantly cheapened rate. In the 1940s and 50s, the influence of Meath farmers was exercised through direct participation in politics at the local or national levels, as county councillors or TDs. The familiarity of large Meath farmers with power and public influence meant that once the IFA was established, its Meath branch quickly developed into a particularly well-organized and powerful element within the organization. However the result was that by the 1970s, Meath farmers had almost completely withdrawn from direct political involvement, preferring to work through the IFA to lobby politicians for support on matters of economic importance to agriculture. By the end of the 1970s, the Meath branch was leading the IFA in arguing that even the Irish state itself was no longer of much importance in relation to policies which affected farming, and that domestic electoral politics should be bypassed in order to enhance lobbying and strategic alliances at the European level (Wilson, 1985). The partial dismantling of the CAP in the early 1990s appears to have done little to change this opinion, as it was widely attributed to international pressure to secure the GATT agreement which the Irish state on its own could do little to fight against, although some commentators are beginning to suggest now that as EU funding for rural areas is switched away from instruments like FEOGA and the CAP, and into Structural Funds for regional development through instruments like LEADER, the withdrawal of large farmers from local politics may turn out to have been a mistake.

The coincidence of the political mobilization of farmers as a collectivity, through appeals to what all members of the profession share in common (and what differentiates them all from other, e.g. 'urban,' interests), with a period of intense pressure from state agents to make farmers accept and adopt a particular, productionist interpretation of what constitutes good farming, has had very significant impacts on how large, modernized farmers like those in Co. Meath have come to define their professional identity. The fact that Meath farmers were becoming less diverse in terms of social, economic and cultural characteristics has been less important in itself than the fact that they were increasingly mobilized around the promotion and protection of an identity which was defined as homogeneous for all 'modern' farmers. A central element of this identity, as it has been shaped particularly by the IFA, is that farming is a business and that farmers must behave as entrepreneurs. This offers some insight into why our interviewees had so little to say about their past, collective or even individual. A significant feature of the entrepreneur identity is that it emphasizes an orientation to the future rather than the past. It stresses that the good farmer is not just skilled in particular ways or the possessor of particular bodies of technical knowledge: more importantly, he (sic) is someone who is always open to new opportunities, someone who is prepared to strike out in new directions, who puts more trust in his own ability to call the odds than in inherited or traditional recipes for success. Meath farmers did not appear to us to think in terms of a 'past' when farmers were not entrepreneurial and a 'present' when they are – rather, they implied that the occupational role of 'farmer' inherently requires entrepreneurial behaviour, so that where such behaviour is absent, we might say, there are no 'farmers,' only landholders.

'Scraping mud is not interesting': views of contemporary farming

From the accounts which our interviewees gave us of what modern professional farming involves, three features stood out. First, they feel that farming has become very heavy in terms of the workload, the routines and clockwatching which it requires; second, it has become a very demanding job in terms of the skills, the technical knowledge and the standards of production expected by the commercial (dairy coops, grain merchants, chemicals salesmen etc.) and state agencies with which farmers have to deal; and third, that the future appears increasingly uncertain, not just for farming in general but for their own farm in particular, as it becomes less likely that any of their children will take over from them, at least full time. There are, however, ways of dealing with the first two of these; and for the third, as long as the option is there to sell up and live on the capital so realized (in many cases half a million pounds or more), farmers who stay in farming are increasingly confronting the idea that they do so because they want to, not because they have some familial obligation to live up to.

One of our interviewees, asked directly what he liked about farming, gave the following reply:

> Well eh I suppose – number one is being your own boss, and eh – the other thing is, the more improvements you make, at the end of the day you can think oh god I've achieved this, you know, you get great satisfaction, you know, job satisfaction.

Most other respondents also emphasized the pleasure they got from doing things well, even while complaining that doing things well nowadays involved a great deal of expense, record keeping and bureaucratic management which they generally did not enjoy. One spoke of his early years in dairying as follows:

> When I started off at first, there was no such thing as recording very much – you know, what breed a cow was, but as we went on then, I started to record, with weight gains and all, and took to keeping a parent book. So you can trace back now, grand-, great-grand daughters, you know, and can remember, you know, even what that cow looked like, you know, her great-grandmother. And you can tell the different milk improvements, so you breed to that – that's what makes it interesting you see, you can't go backwards. Well if you do, you're out of the business.

Another, in the middle of complaining about the trouble and expense he has endured in intensifying his cattle production in recent years, particularly building them slatted sheds to pass the winter in, added:

> I suppose the more you get intensive, the more money you make and the more you can spend to make life comfortable. That's the way I would look at farming. I certainly wouldn't farm today the way I came into farming. I think it was drudgery really – mucking out the cattle every day, milking by hand, you mightn't work that hard but I thought there was a lot of drudgery attached to it. Boring work, no money in it. Today machines do most of that kind of work, if you can afford to have them put in. I mean if you have slats you can have scrapers, it's all done [mucking out] by scrapers. You're just constantly feeding cattle, looking after cattle, which is interesting. Scraping mud is not interesting . . .

One farmer, who has a 450-hectare farm given over to rearing suckling calves and running sheep, spoke for many of the others when he said that today:

> We seem to be running faster and faster just to stay in the same place . . . We're not on the breadline, like, we've a nice, a reasonable income, but for the amount of hours that you [addressing his wife, a secondary school teacher] work and I work, we're not . . . I'm not moaning about it, this is what we wanted to do, we made a conscious decision to be in farming, but you get fed up sometimes. For example, I'd sometimes like to be able to come in at 5 o'clock on a Friday evening and say, that's grand, I don't have to think about that till Monday morning.

However, many of the farmers we spoke to were precisely starting to think of their work like that – starting in fact to assimilate farming into other professional work models in which a clear differentiation is made between work time and leisure time. As one large cattle producer told us,

> The thing to do is to get organized. Now I don't mind what other people do, but my main work here should be finished, we'll say about 6 o'clock if I'm organized – with the exception of silage or, you know, something like that

The same man was proud of the fact that he and his wife always took a foreign holiday once a year, no matter what, and that he could spend a lot of leisure time on his hobby, playing golf. He was very critical of

> dairy farmers who do nothing but work, and at the end of the day they're no better off. If you rest, you can do more work inside the same hours than if you work continuously, you know.

Many farmers said they felt under pressure because of the lack of affordable, or available, employed labour nowadays, and the lack of help from their children who were much more engrossed in schooling than used to be the case. But as one of the bigger dairy farmers pointed out:

> You have always got to work hard in farming, but the only thing I can say is, in dairying, you certainly can afford to pay a little more labour, so from that point of view you could well say, maybe, you haven't to work so hard. Because the margins are so small in dry stock and tillage that you can nearly hardly pay labour, whereas in dairying you can carry some labour.

He suggested, as other respondents also did, that if farmers had a choice about the commodity line they wanted to get into, considerations of leisure time and of being able to get some paid help were becoming increasingly important to them in making their choice.

The farms which we studied were likely to have gone through constant processes of change and development over the farmer's life history. While the farms tend to be fairly specialized in one enterprise of another, they also have a history of changing specializations at different times. These changes were obviously in part attempts to settle on an enterprise system which seemed economically sound in a given period, and which also seemed well suited to natural features of the farm; but we were also surprised by the extent to which a liking for and interest in certain sorts of work – rearing animals, growing crops – on the part of the farmer or his intended heir influenced decisions to change. The following account of his farm's history by one interviewee may stand as a typical example. His memory of his farm went back to the 1950s when his father was running it. In the 1950s they were 'reasonably big' into dairying – perhaps twelve to fifteen cows, all milked by hand:

> There was a good lot of money in it in 1957 – it was easy to keep a house like. It was a regular income, and most of the stuff they would have anyway. They bought nothing – they kept a good garden. Most people did in those years – no-one keeps one now, you buy your vegetables . . .

He himself came into farming in the mid- to late-1960s, at which point his father had increased the dairy herd to around twenty cows and also started keeping some sheep. His own early moves were towards increasing the dairy side of the farm, mechanizing milking and moving into a more intensive system of feeding. Between his father and himself, over the period since the 1950s, they continuously added to the size of the farm: his father started with around 90 hectares, he inherited around 160 hectares, and he now owns about 500 hectares and rents a good deal more. His latest land acquisition was in 1993 and was tillage land, with no milk quota on it. But he is not interested in adding tillage to his operations, rather he has been 'drifting towards dry stock' which at this stage in his life he prefers to dairying. He plans to use his new land for holding stock. However, even this is only for the moment:

> I'll stay where I am at the present time because I have enough to do and it depends on whether the young lads has any interest in it or not – the next generation – if they have I may try and set them up. But I'm not going to strangle myself either, but I may sell that farm now and buy a farm with a quota, if there's any of them interested in dairying.

Such comments show that liking the work, exercising skills, and ability to organize acceptable work routines, are increasingly seen as central features of modern farming. Farmers want their work to be interesting; and many farmers defined 'interesting work' as work which demanded some degree of technical or mechanical knowledge. When it came to keeping a son on the farm to inherit, we were frequently told that sons like machinery: the reason why a farm might have more tractors, for example, than it realistically needed was because that gave the son opportunities to learn and exercise mechanical skills. Others suggested that farmers like to feel they can make their own decisions about such things as chemicals use, and not be totally dependent on advice from the chemical company or the state advisory service. As one intensive tillage farmer said:

> The farmer's a very cute man when it comes to money so he'll always use a bit less [than the prescribed dosage of e.g. weedkiller] . . . The actual chemical usage round the farm, there are restrictions and guidelines there that we are supposed to adhere to, but they are not inspected properly . . . I've never had anyone come round here and inspect my yard. So it's left up to the individual farmer and em, I think our

> record is quite good . . . But you'd be surprised, particularly with the tillage farmers, how well up they are on, on the actual products and the use of them. We can all use calculators now . . .

In summary, a major emphasis which came through Meath farmers' talk about being a farmer was that farming nowadays is all about control and change – about being able to cope with, adjust to and as far as possible control a rapidly changing external environment. Being good at that – which included on occasion finding out how to beat the authorities who were forcing the changes through – was part of the skills which marked out a good farmer. The person who stopped being interested in developing his enterprise, diversifying if he thought it looked promising, or trying to beat the system when it merited it, was seen as someone who had virtually retired from being a professional farmer.

'So if he doesn't want it, I'm not going to press him into it': occupational succession in Meath farming

'Family farming remains the most hereditary of professions' (Potter and Lobley, 1996, p. 286). In fact, one reason for arguing that farming is still un- or under-professionalized is the connection which continues to exist between entry into the occupation and inheritance of family property. It is still the case that the great majority of Irish farmers come into farming through inheriting a farm, and even those new entrants into farming who have bought rather than inherited land often come from farming backgrounds. However, Meath farmers talked about the issue of family succession on the farm in ways which suggested that they no longer see the connection as automatic. In a number of ways they were beginning to detach their own involvement in and sense of pleasure from the successful practice of farming from an understanding of this as something which will pass on to the next generation of their family.

The majority of the farmers we interviewed were in middle age (mid-40s to mid-50s), at a stage in the family life cycle where they had teen-aged children who were starting to think about their future careers. A few were a good deal older, and in the process of either gradually ceding the reins to an inheriting son or beginning slowly to 'retire,' that is, scale down their involvement in the farm because they had no successor. Of those who had potential heirs still at school, virtually all expected that their heir would go on to agricultural college and get the proper professional accreditation for the occupation, even though they themselves had generally not done this and, indeed, many of our interviewees had left school at a very early age and without any official education qualifications at all. Daughters generally still did not count as prospective heirs, even when they displayed an interest in the farm which the sons did not. Although it is not in any way 'traditional,' Meath farming is still quite patriarchal.

In other less professionalized forms of farming, planning for the eventual takeover of the farm by a son is a major, and sometimes perhaps the only, way in which future-orientation in organizing the farm occurs. This is still very salient for Meath farmers too, many of whom are still hopeful that they will be able to pass on the farm within the family. As mentioned earlier, some were trying to develop the farm operations, technically or mechanically, in ways which they thought would make it more attractive to the next generation (this 'succession effect' on farm development has been noted in a number of other studies of family farming, most recently by Potter and Lobley in the piece cited above). In other cases, farmers bought in more land on which prospective heirs could experiment to see how they took to the occupation. One interviewee, for example, had bought a piece of land across the road from his main farm which he was using to let his son try out his skills in farming. Asked if it has always been more or less taken for granted that that son would continue to farm, he replied:

> Oh it is – the older son didn't want to farm. So I never asked him to stay at home. I always think you should never force a young fellow to farm, of all the jobs going. You either like it or you don't – and if you don't like it, you're not going to make a go of it. My elder son is in the States . . . I've one daughter, she's working in Navan [the county capital], she's the youngest of the three.

And a number of the dairy farmers we spoke to said they were staying in dairying, even when they personally would quite like to get out of it, because they thought the guarantee it provides of a monthly pay cheque makes it more like an urban job, which would make it more attractive for a son to take over from them.

Others were reluctantly coming to terms with the fact that they would probably have no successor. In one interview, for example, the farm wife said they were thinking of selling out, on the grounds that:

> Nobody wants it here, certainly the two girls, one of them wouldn't set foot outside in the yard, the other one does it when she has to do so.

Her husband then broke in on her, saying:

> When my father bought this farm, he came from a *very* small farm, and himself and my mother bought this farm in 1938, so he worked all his life, and then when it came to me, I had two older brothers and both of them went off to their own careers, so, you feel you have a duty to pass it on to the next lad, you know? So if

> he doesn't want it, I'm not going to press him into it, I'd *like* perhaps if he'd want to farm, I don't think I'd envy him because I think it's a tough life, I'd much prefer him I think to have some other string to his bow, some alternative and in fact we 'll make sure that he does, and maybe do it besides the farm, and that's what a lot of farming is going to be in the future, part-time farming. It's a terrible attitude to be developing in young farmers, but that's what's happening all over

However, interspersed among these farmers were others whose comments suggested that they were oriented to the future in quite a different way. Their remarks suggested that they saw no particular reason to expect that the son of a farmer would himself become a farmer, unless he had a particular aptitude for it – even, indeed, that the fact that the father's profession was farming was not really the children's business at all, except perhaps in the same way as the family of a doctor in general practice may be affected by his professional routines and duties. One large tillage farmer with teenaged children, for example, said he thought his children rather envied his way of life, but that was because they only saw the good things about it, the freedom from bosses and the chance to spend time in the open air, and were shielded from the less attractive sides. He thought one son might become a farmer, though probably not on the farm they were on at present nor in the same commodity line as his father, but this was mainly because he did not seem likely to find success at any other job:

> If the eldest goes to university which I hope he will do I would hope he will be OK and have, em, a proper career. The others, one of them is extremely dyslexic – very hardworking, very clever – and eh, I would like to think that I would possibly set him up in some farm enterprise, whether it's horses or whatever. But basically no, I would direct them all into some other profession because this place wouldn't sustain two families, and em, I'm possibly going to sell it anyway . . .

Meath farmers, increasingly, see themselves as part of a national and even, in some cases, international bourgeoisie. For many of them, it appears, their main concern about their children is that they too should find places within the national bourgeoisie – but not particularly within the local agrarian bourgeoisie of Meath.

'It's nature': productionism, professionalism, and sustainable agriculture

At the start of this chapter it was suggested that the current interest in rediscovering and revitalizing 'traditional' farm cultures is at least in part linked to a belief that this is an important way of achieving greater environmental sustainability in agriculture. Making farming more concerned about sustaining as well as exploiting nature has been seen, in this perspective, as requiring a reversal of the impact of productionism on a supposed earlier culture of farming which stressed landstewardship and avoidance of over-exploitation of nature. The discussion of 'productionist' farming in Meath however suggests that such a culture may not have existed at all in any locally distinctive form, or that if it did, it has no relevance at all for contemporary farming practices or for the way in which farmers in the area understand farming. This is not meant to imply that Meath farmers are not and have never been concerned about sustainability in their use of land and nature. It is rather to suggest that instead of looking to the past we look for occasions *within the present discourse* of farming where concerns about sustainable farming may be located or could be introduced. This might allow us to move towards a position where, instead of seeing productionist orientations to farming as completely irreconcilable with concern about the environment, we could think of it as something which occurs in a number of different social, economic and political contexts, some of which are much more hospitable to the expression of such concern than others. It might also help us to move away from that over-focus on urban constructions about rural environmental attitudes which Miller complains of in British rural sociology and become more open to incorporating farmers themselves as agents of environmental management into the development of policies for a sustainable agriculture (Tovey, 1994).

Our research on Meath farmers allowed us to identify at least two points where considerations other than those of maintaining very high output levels and intensifying inputs are evident in what they say. One was in the way they talked about recent policy changes intended to reduce productionism; the second in the way they talked about 'nature,' including farmed nature.

Meath farmers are, as might be expected, very aware of recent changes in direction in EU agricultural policy. They recognize that, as one farmer put it, the days when 'you could produce everything you want and as much as you could of it' are gone and controls on production are now the order of the day in nearly every commodity line. Most could see that shifting over to a more extensive type of farming system might have benefits, for the environment and also for farmers themselves if it meant more leisure time or less indebtedness. But it was not an option they expected to be widely taken up. Asked if he thought it was a good idea, one dairy farmer replied:

> Well I suppose – take tillage people – set-aside. They'll push to produce as much corn on less land. They'll push. We've pushed, we've pushed the milk yields. People

> thought dairying would collapse with quotas. It has done a good thing for dairying. We're producing more milk now with less cows than we did six, seven years ago. We're producing it more efficiently now . . . High-cost producers were making more money per cow than low-cost producers. If you put in low cost you come out with low yields – net profit per cow – whoever puts in most and has the highest yield has the biggest margin in the end. I'll stick with that . . . And I think that's what will happen in tillage – they'll still produce as much corn on less acres . . .

Comments by a tillage farmer who concentrates on growing winter wheat and who experienced having to set aside 15 per cent of his land during the winter of 1993/4 support this prediction:

> I think I can still produce as much yield, off the [remaining] acres managed properly, you know, and even enthusiastically – in fact [laughs] it's my full intention to try and do it, the smaller area you have the more output you get out of it, you're just doing things that little bit better . . .

Meath farmers still clearly think of agriculture as a business of producing food, and believe that the more you produce of it the better the business you are running. However it became apparent that this was not always just a matter of commercial sense, but actually of morality. One farmer, who has no tillage on his own farm, made comments about the policy of set-aside which seemed typical of many of those interviewed:

> I don't think it is right at all. It's against the farmer's nature. The land is there to produce food. We're producing too much now, I suppose, but a lot of the world still can't buy food. I suppose that's the problem. It's very frustrating. When you see land – 20, 30 acres just growing weeds – the best of fertile land . . . I like the land. It's there to feed the workers, the population. It should be used. We shouldn't have starvation in the world, it's terrible.

This concern about poverty is not just rhetoric – the IFA has for some years been developing linkages with farmers in Third-World localities, particularly in Africa, through which Irish farmers send gifts of heifers-in-calf for milk production, new breeds of calf, appropriate machinery and what they can themselves contribute in the form of farming skills and experience. For those involved, the idea that good land should be made to lie fallow when it could be producing food for hungry people is a moral affront. Although the IFA have not articulated the issue in this way for them, some of our interviewees seemed to be moving towards the claim that environmentalist concern about over-exploitation and pollution of nature in the First World should be balanced by concern about continuing food shortages in the rest of the world.

At the heart of a professional self-image which is rooted in the role of producer, then, elements of a non-productionist, moral discourse can be detected. Similarly, elements of aesthetic and emotional orientations to nature are embedded in these farmers' attitudes to their occupation, particularly in the part played in their working lives by their likes and dislikes of different enterprise lines. Most were forthright about these, describing themselves as either a 'stock man' or a 'tillage man,' and within the 'stock' category as either someone who liked dealing with cows and young calves, or someone who liked finishing beef cattle (the latter tended to hold a disparaging view of heifers and cows whose meat is generally less highly rated and often kept 'just' for the domestic market). Just as you cannot force a son to follow you into farming if he doesn't like the work, they suggested, equally you yourself will not be a good farmer if you are in the wrong enterprise category for your personal taste.

The extent to which these farmers were emotionally involved in their work and were prepared to talk about their emotional responses to it was to us surprising. Aesthetic and appreciative orientations are embedded alongside rationally calculative ones in their daily working lives. Dairy farmers, for example, often mentioned how much they liked cows. One farmer, who since his wife's death has been occupied more in the house and has turned over most of the milking to his son, said regretfully:

> The only way if you really want to have really close contact with your cows is to milk them. You know, I'd go out in the yard [nowadays] and I'd know some of them, but not like I used to. When I was milking I could nearly know a cow by feel. You know, and they're very intelligent animals too – nearly the same ten or fifteen animals would always come in first. You know, they'd nearly know when to come in . . . The young ones would always be last because they'd be the babies in the family, the young heifers . . .

Another farmer, asked why he was fond of his animals, replied:

> It's nature – you're probably there for the birth of most of them – it's nature at its best to see how they turn out. Some would be lost on the way and you'd feel lonesome. I'd feel lonesome for cows now, they're with you six, seven, ten years and you'd hate to see them dying from something all of a sudden.

And a tillage farmer said, if rather embarrassedly, that one of the reasons he was in tillage himself, rather than cattle or sheep, was that he would not be able to face sending animals he had reared and grown fond of to be slaughtered.

Domestic animals, especially dogs and cows, can come to be regarded by their owners as part of the social system of the farm. Thus cows make up a 'family' which is characterized by social relations of hierarchy, similar if inferior to the farmer's own family; and farmers can feel a strong, paternalistic sense of responsibility for their well-being. In this way of talking about 'nature' – in a sense, portraying it as a lower level of being human – farmers were combining their professional orientation towards production and output with a version of 'husbandry' or 'stewardship,' rather than letting one supplant or repress the other. Crops were not socialized in the same way, but farmers' attitudes to them were replete with aesthetic values. Tillage farmers spoke of standing looking at a 'clean' (weed- and disease-free) field of corn in growth, or how they enjoyed the sensation of feeling the grain at harvesting:

> To put your hand into the grain bin on the combine and see the quality, it's fantastic . . . When you see em, lovely plump colourful grains, you know you've done your job properly and it's very satisfying.

Farming as an occupation, as noted earlier, is thought by farmers to provide exceptionally high levels of job satisfaction and this can apparently derive from the aesthetic and emotional as much as from the practical achievements of production. In fact it is almost impossible to separate the two; as MacEachern argues in her study of sheep farmers in Yorkshire, 'When we look at what farmers, here dales farmers, were doing and saying in their everyday lives, there was a blurring of the divisions between business and conservation . . . The exploitation and stewardship inputs of farming actually take particular form and meaning in the very work which farmers do, meanings which may be at variance with other understandings of the same input' (1992, pp.162–3). To try to allocate one to a contemporary productionist culture of farming while treating the other as a residue from an earlier traditional culture is to do violence to how farmers experience their own work.

Farmers from all commodity categories, finally, commented spontaneously on their concerns about farm pollution. This concern had apparently arisen out of the campaign conducted by the IFA since the late 1980s to alert farmers to pollution dangers, especially pollution of watercourses by silage or slurry effluents. We might conclude that farmers have accepted that they have a responsibility to keep the countryside unpolluted, not because they have themselves been converted from 'productionist' to more ecological orientations but because it has been important to the IFA to take the lead in reducing agricultural pollution in order not to let control over the countryside slip into the hands of non-farming or urban groups, such as environmental scientists or landscape managers. This is not to suggest that farmers are not genuinely concerned about the effects of their activities on wildlife or the landscape, but that the salience of these concerns to them is influenced by the broader context. The same circumstances under which the contemporary conception of professional farming has arisen among Meath farmers – in particular their availability for mobilization by the state and the IFA around an ideology of output and entrepreneurship – are capable also of bringing these farmers to recognize that caring for nature is part of professional behaviour too.

Acknowledgements

The material on which this Chapter is based was collected as part of the Irish contribution to the 1993–1995 cross-national study *Land, culture and crisis – From productionist success to fiscal and environmental impasse on European farms 1940–1990,* funded under the EU SEER Program, Area III, and coordinated by Dr. Simon Miller of the Centre for Rural Research and Policy, University of Manchester, England. The Research Assistant on the project, without whose commitment it could not have proceeded, was Geraldine Haverty.

References

Anderson, B. (1983), *Imagined Communities: Reflections on the Origins and Spread of Nationalism*, Verso: London.

Arensberg, T. and Kimball, S. (1940), *Family and Community in Ireland*, Harvard University Press: Cambridge.

Connerton, P. (1989), *How Societies Remember*, Cambridge University Press: Cambridge.

Curtin, C. (1986), 'The Peasant Family Farm and Commoditisation in the West of Ireland,' in Long, N. et al., *The Commoditisation Debate*, Wageningen Agricultural University: Wageningen, The Netherlands.

Frank, A.G. (1969), 'The Sociology of Development and the Underdevelopment of Sociology,' in *Latin America: Underdevelopment or Revolution?* Monthly Review Press: New York.

Giddens, A. (1991), *Modernity and Self-Identity*, Polity Press: Cambridge.

De Haan, H.J. (1995), 'The Political Mobilization of Tradition and Reflexive Modernization,' paper presented to the XVI Congress of the European Society for Rural Sociology, Prague.

Hannan, D.F. (1979), *Displacement and Development – Class, Kinship and Social Change in Irish Rural Communities*, Economic and Social Research Institute: Dublin.

Johnston, R.J. (1991), *A Question of Place*, Blackwell: Oxford.

Jones, D.S. (1983), 'The Cleavage Between Graziers and Peasants in the Land Struggle, 1890–1910,' in S. Clark and Donnelly, J.S. Jr. (eds), *Irish Peasants, Violence and Political Unrest*, Gill and MacMillan: Dublin.

McEachern, C. (1992), 'Farmers and Conservation: Conflict and Accommodation in Farming Politics,' *Journal of Rural Studies*, Vol. 8, No. 2.

Miller, S. (1995a), Report on the UK Case Study for the project *Land, Culture and Crisis: From Productionist Success to Fiscal and Environmental Impasse on European farms, 1940–1990* (unpublished), for EU DG XII.

Miller, S. (1995b), 'Land, Landscape and the Question of Culture: English Urban Hegemony and Research Needs,' *Journal of Historical Sociology*, Vol. 8, No. 1.

Murdoch, J. (1994), 'Sustainable Knowledge,' *Geoforum*, Vol. 25, No. 2.

Potter, C. and Lobley, M. (1996), 'Unbroken Threads? Succession and its Effects on Family Farms in Britain,' *Sociologia Ruralis*, Vol. 36, No. 3.

Reidy, K. (1992), 'Agriculture and the Countryside (Findings from a Survey of Farmers' Attitudes to Selected Issues),' *Proceedings of the Agricultural Economics Society of Ireland 1988/89.*

Schutz, A. (1971), 'Choosing among Projects of Action,' in *Collected Papers Vol. I, The Problem of Social Reality*, Martinus Nijhoff: the Hague

Symes, D. (1992), 'Agriculture, the State and Rural Society in Europe: Trends and Issues,' *Sociologial Ruralis*, Vol. 32, Nos 2/3.

Tovey, H. (1992), 'Rural Sociology in Ireland – a Review,' *Irish Journal of Sociology*, Vol. 2.

Tovey, H. (1994), 'Rural Management, Public Discourses and the Farmer as Environmental Actor,' in Symes, D. and Jansen, A.J. (eds), *Agricultural Restructuring and Rural Change in Europe*, Wageningen Agricultural University: Wageningen, The Netherlands.

Van der Ploeg, J.D. (1992), 'The Reconstitution of Locality: Technology and Labour in Modern Agriculture,' in Marsden, T. et al. (eds), *Labour and Locality – Uneven Development and the Rural Labour Process*, David Fulton: London.

Wilson, T. (1985): *Beef Before the Ballot (The Impact of the Common Market Policy on Agriculture and Politics in Eastern Ireland)*, Unpublished Ph.D. dissertation, Dept. of Anthropology, City University of New York.

Wilson, T. (1988), 'Culture and Class Among the "Large" Farmers of Eastern Ireland,' *American Ethnologist*, Vol. 15, No. 4.

Zonabend, F. (1984), *The Enduring Memory (Time and History in a French Village)*, translated by Anthony Forster, Manchester University Press: Manchester.

8 Agricultural change and innovation groups in the Netherlands

Ina Horlings

Introduction

Dutch agriculture faces the challenge of sustainable development. The report *Our common future* defines sustainability as a process of change in which the utilization of resources, the direction of investments, the orientation of technological and institutional change are all in harmony with each other and (all) increase both current and future possibilities to meet human needs and desires (WCED, 1987). In order to start off sustainable agricultural development, the government has formulated several environmental measures in strategic policy documents (Ministerie van VROM, 1989, 1990, 1993; Ministerie van LNV, 1990).

Farmers who endorse more sustainable agriculture often feel, however, that government policy does not give them an incentive to produce in a more ecologically sound way. They feel that the government restricts them more than necessary and exerts too much control over their operational management. In addition, they feel 'trapped' because of the combination of deteriorating economic conditions and environmental policy measures. They are forced to change their management within narrowing operational margins.

The key hypothesis in this Chapter is that farmers are hampered to achieve more sustainable agriculture by the 'farm context.' This context leaves them less freedom to respond in a differentiated way to ecological problems. The farm context is the network of governments, organizations and institutions that is relevant to farm management. The interaction between farmers and this organizational context has led to a narrowing operational space, which limits the possibilities for farmers to develop their own solutions to ecological problems based on their skills and knowledge. Differentiated agriculture and craftsmanship are considered in this Chapter as necessary but not sufficient conditions for a more sustainable agriculture.

However, this is only half of the story. The influence of the farm context should not be understood in a deterministic way. Farmers interact with actors in the farm context, and apply advice and services selectively, suited to their own specific situation. One way in which farmers react to agricultural developments is by creating 'innovation groups,' which could contribute to more sustainable agriculture. Innovation groups are dissatisfied with developments within agriculture and react by trying to enlarge their operational space and finding their own ways of solving ecological problems. Based on their motivations, goals and actions these groups can be characterized as 'social movements.'

This Chapter will provide answers to the following questions:

- what are the main changes in Dutch agriculture during recent decades?
- what are the consequences of these developments for Dutch farmers?
- how do Dutch agricultural innovation groups react to these consequences?
- how do Dutch innovation groups contribute to more sustainable agriculture?

Changes in agricultural production

This paragraph and the next describe the main changes in Dutch agriculture during recent decades. Agricultural development in the Netherlands has led to severe ecological problems. These problems are not only caused by changes in production, but also by social changes and changes in the relations between farmers and other actors. As a result of these, the possibilities for farmers to develop more regionally based sustainable agriculture based on craftsmanship have diminished (Figure 8.1).

Structural factors like technological and institutional innovations have had a strong influence on the character and direction of agricultural development. The relative costs and scarcity of capital, labour and land were very important determining factors (Hayami and Ruttan, 1985). In the Netherlands land is scarce and land prices are high compared to production-raising technology or labour-saving machinery. As a result, the main development trust in Dutch agriculture was intensification of production.

The development of Dutch agriculture has moreover been strongly influenced by exogenous factors such as external knowledge, capital and technology. Especially in respect of cultivation under glass and factory farming, production systems have originated that are to a large extent determined by exogenous

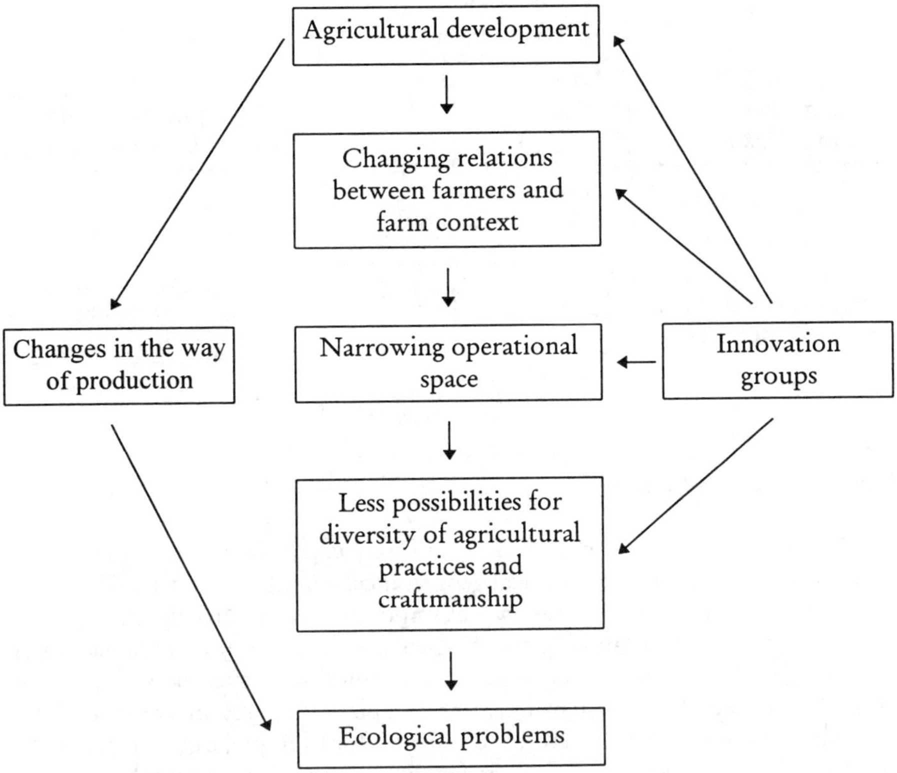

Figure 8.1: Changes in Dutch agriculture

factors. Expansion of markets, partly due to the establishment of the common European market and related industrialization of production, were of vital importance in this respect (van Dijk, 1994).

The government played a favourable and supportive role in agricultural development through stimulation of research, information and education and technological innovation and by investment in infrastructure. The effect of government policy on the structural adaption of the sector and on research and development contributed to the fact that The Netherlands has become the most capital- and knowledge-intensive agriculture of the world.

Technological development has made it possible for agriculture to function more and more independently of specific local conditions, which had previously been the foremost determining factors for the production system. Van der Ploeg

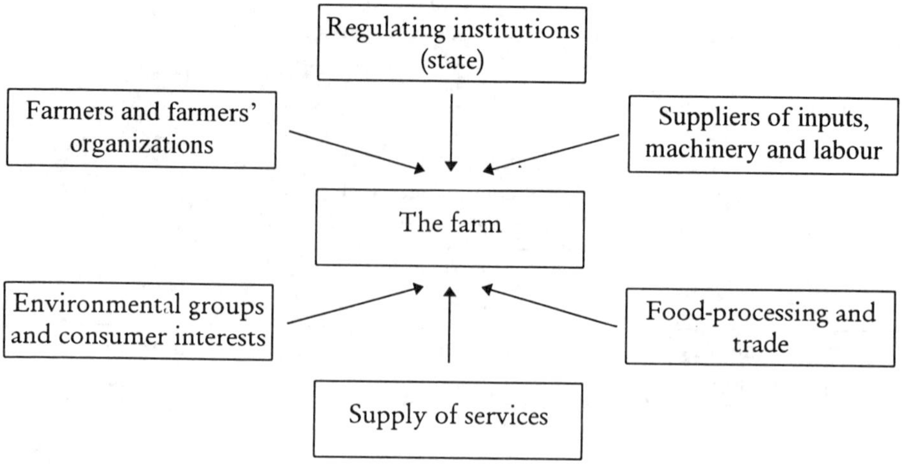

Figure 8.2: The farm context

(1992) describes this as the 'unlinking' of agriculture from local conditions. 'Locality' not only includes physical-geographical conditions, but also the social structure of a region. I will give two examples of this 'unlinking' process.

Firstly, there is an unlinking of ecological processes. To lessen the insecurity that is related to natural processes, natural conditions such as soil fertility, temperature, availability of light and mineral supply are increasingly manipulated in agricultural production. Natural conditions, for example, can be replaced by synthetic 'growth regulators' such as fertilizer or lighting in greenhouses.

A second example is the unlinking of specific end products. Agriculture mainly results in raw products and increasingly less in end products. This enlarges the distance between producers and consumers. The quality of the product used to be linked with the farmer's labour process, the way the farmer worked. However, quality is more and more an aspect that is added in the form of artificial odours and flavourings at the end of the production chain. At the same time quality is constructed by images and recommendations to consumers, which refer to 'greener,' 'more natural,' or 'healthier' products (van der Ploeg, 1990). Agricultural products are increasingly broken down into exchangeable food components. This leads to a further de-regionalization of food production and consumption. It is technically possible to make 'fake' local products like feta (sheep's milk cheese) from cow's milk in the Netherlands and 'French camembert' in Germany (Jongerden and Ruivenkamp, 1996).

These developments in Dutch agriculture have had consequences for production and the physical environment. What is most striking is the steep increase of productivity and intensification through the use of fertilizers; refinement and crop protection. Increasing demand for land has stimulated intensive industrial farming systems which do not use land (de Hoogh, 1985/86) and there has been a strong tendency towards specialization and standardization.

These have had negative effects on the environment, nature and landscape. In terms of fertilizer and pesticide use and the number of livestock per hectare, this environmental pressure in the Netherlands is the highest in Europe. High energy consumption, especially through cultivation under glass and the production of waste and emission of greenhouse gases which contribute to climatic change, are other examples of environmental pressure. Furthermore, agriculture has a negative influence on nature through lowering the water table, parcelling out of nature reserves and reduction of biodiversity.

Social changes in Dutch agriculture

Agricultural development also had social consequences. These resulted from the increasing influence of the 'farm context' – the 'organizational network' with which farmers are in technical-administrative contact. Benvenuti (1982) referred to this as TATE (Technological-Administrative Task Environment). The farm context includes all sorts of organizations and institutions such as consumers, suppliers of inputs (like fertilizer and pesticides), capital and labour, a variety of institutions such as (semi)-authorities, research institutions and the media (Benvenuti, 1991) (Figure 8.2).

The organizations within the farm context maintain not only economic but also social relations with the farm. The actors in this context supply the farmer with products and services, but simultaneously confront him with specific demands and regulations, for instance, the regulations from the agricultural advisory service concerning the use of crop protection substances or the obligations connected with contract farming.

Besides the farm context, it is useful to distinguish the wider 'agricultural context' (Figure 8.3), which is beyond the local or regional level. The import of raw materials for fodder from Third World countries and the export of milk powder imply, for instance, that geographical aspects of agricultural networks should be taken into account. The agricultural context also covers relations that do not affect farm management directly, but indirectly influence the character and direction of agricultural development. Examples are social norms and values, opinions about the relation between man and nature, industrial development of rural areas and international economical developments.

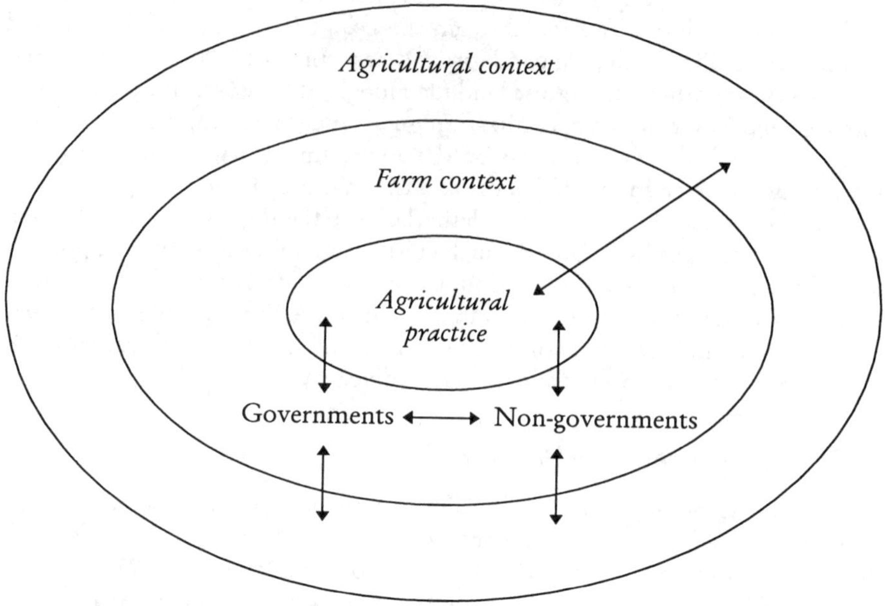

Figure 8.3: Relations within the farm context and the agricultural context

Characteristic of the relation between the farmer and the farm context is that it is asymmetric, which means that the farmer depends on the supply of services, such as credit from a bank or management expertise and advice. Control over farmers is exerted by stimulating certain investments and the promotion of a specific model of agricultural development. Relations between farmers and actors in the farm context have also become more and more formal. Farmers must obey the laws resulting from (environmental) policy and they must follow the contracts agreed upon with agro-industrial firms. The social consequences of the specific nature of the relations between farmers and the farm context are an externalization of farmers' tasks, scientification of knowledge and standardization of farm management.

The standardizing influence exerted by external organizations threatens the diversity of farming practices and the survival of specific farming craftsmanship. Standardization has decreased operational space for reacting to external influences in a variety of ways. This is how the farm context has in fact hampered farmers from testing different innovations in the field of sustainability. Authorities and the agricultural advisory service, for example, used to stimulate a certain type of farm that was strongly orientated to the implementation of the latest technology and characterized by large-scale, intensive production, the so-called 'leading farm.' When striving for less environmentally damaging agriculture this can be considered a problem. The 'leading farms' are not automatically the farms with the least pollution or best-equipped for anticipating environmental regulations (Roep et al., 1991).

Diversity of agricultural practices is important because it offers the possibility for exploring potential directions towards sustainable farming. Reducing the choice to a single model of agricultural development does not do any justice to this learning process and bears the risk of not being 'tolerant of errors.' This means that, for example, because of heavy capital investments (such as industrial processing of manure) it can be hard to change direction when the technology produces negative effects for the farmers or the environment. In plain words, diversity, if directed to sustainability, can raise more possibilities for renewal.

The second result of the diminishing operational space of farmers is the danger of less importance being attached to 'farming craftsmanship.' Farming craftsmanship includes two aspects, handicraft, which refers to the link between brain and manual work, and farming knowledge or experience. The result of craftsmanship is sometimes called *l'art de la localité*, the art of the specific, building on experience and knowledge arising from specific conditions (van der Ploeg, 1991). A typical example of farming craftsmanship and the relation with local conditions is the production of quality wine in France, in which the local conditions (the *terroir*) can be tasted in the wine. Farming craftsmanship arises as a result of a continuous learning process.

The influence of 'off-farm' knowledge from scientific and other research institutes is growing. However, craftsmanship is still the means by which the farmer selects the kind of scientific knowledge which best fits his management; it filters and transforms external knowledge. In a situation where farming experience is undervalued and is replaced by scientific knowledge, farmwork is transformed into one standard model which leaves little space for variations. Local knowledge systems, however, can function as a corrective mechanism for externally developed knowledge. Craftsmanship is the leading power behind farming renewal and can be used to advantage in achieving a more environmentally sound agriculture.

In the past, farmers used this craftsmanship for learning to deal with sprays, fertilizers and technology and to increase their production constantly. Sustainable agriculture demands development of new knowledge and skills, for example aimed at mechanical weed control, quality production and nature and landscape conservation. For this, farming craftsmanship is a necessary condition. For example, to promote valued natural elements such as rare birds and plants on intensive farms, specialized skills and knowledge are required. Without wishing to idealize farming craftsmanship, not using this reservoir of knowledge to stimulate farmers in the direction of sustainability, is a waste of creativity.

In conclusion, as a result of the direction of agricultural development farmers are faced with fewer possibilities of applying different methods and using craftsmanship to find their own routes towards sustainable agriculture. Agricultural work has been de-professionalized: farming is being reduced to mass production and the supply of raw materials for industrial production, unconnected to natural conditions (soil, water, climate) and divorced from contact with consumers. Care of nature and landscape is taken over by specialized organizations and the importance of craftsmanship as the basis of agricultural knowledge is increasingly replaced by scientific knowledge. Growing international competition for markets between agricultural producers has, moreover, exerted pressure on the cooperation between farmers.

Agricultural development and agricultural diversity, a paradox?

The foregoing pictured the farmer as a victim of the influences from the farm and agricultural context. However, this is only half of the truth. Agricultural development is not a linear, irreversible process. Counter-forces in agricultural practice prevent farmers from becoming passive executers of orders given by the farm context. Farmers do not passively wait for their marginalization and agricultural development does not necessarily lead farmers to total dependence.

Agricultural development cannot be understood only by analysing external influences. The mutual interaction of internal and external factors should be taken into account as well. The 'actor approach' offers such an analysis; it focuses on actors' different reactions to similar circumstances (van der Ploeg, 1991; Long, 1989). Farmers are knowledgeable and capable of influencing their own circumstances, regardless of their power. They have the capacity, though their operating space is narrow, to make decisions and to innovate and experiment (Long, 1989). The actor approach takes the characteristics of farmers and farming into account. Characteristic for agriculture is the way farmers are dealing with all sorts of risks. Specific to land-based agriculture is the uncertainty of ecological and climatological conditions. In short, farmers are confronted with the following insecurities (Wolpert, 1964):

- personal factors (health of the entrepreneur, the family and employees);
- institutional regulations (for example, government policy concerning prices of inputs and outputs and statutory regulations for environmental investments);

- technological changes and market structures;
- physical factors (weather).

Economic-technological and socio-political factors have played an important role in agricultural development. However, in spite of the processes of homogenization and standardization, agricultural practices show a great deal of variety and differentiation. For this paradox, the following explanations can be offered.

- Not only exogenous factors but also factors inside and outside agriculture and horticulture, endogenous and exogenous factors, determine the development of the agricultural sector. Development patterns, exogenous or endogenous, are always influenced by local factors like knowledge and labour. Local sources, like farmers' knowledge, are influenced by exogenous elements like education institutes and like media. Endogenous and exogenous development are therefore relative concepts. In order to understand agricultural development, not only the sources that influence the development should be taken into account, but also sources that control farm management: farmers, farmers' groups or other actors. In other words, are farmers the object or the subject of agricultural development? (Lowe and Murdoch, 1995). Endogenous agricultural development is defined here as mainly, but not exclusively, based on locally available resources, mainly controlled by farmers (groups) themselves. In Dutch agriculture, the influence of exogenous factors has been transformed by endogenous factors. Farmers work in different regional contexts, with variable soils, nature, landscapes and groundwater levels. External technology and knowledge are adapted to these local circumstances. An example is the adaption of manure injection machines to the weak peat soils in the West of the Netherlands.
- External innovations are not everywhere, at the same time and in the same way, appropriate for all farms. This can be explained by differences in social characteristics, demographic structure and culture. For example, intensification in the south-eastern part of the country was stimulated by a pattern of succession that guaranteed each child an equal part of the land. In some north-eastern parts of the country the land was inherited by only one of the children, which led to less scarcity of land (de Haan 1994).
- Farmers have different opinions and react in different ways to external influences. The manner in which farmers react to the insecurities they face does not only depend on their knowledge, but also on their opinions. The different responses of farmers to developments in markets, technology and policy lead to different strategies or 'styles of farming.' Each style of farming reflects the farmer's view of 'good farming.' A style is visible in the way a

farmer breeds his cattle, the management of pasture land, the use of off-farm capital, the use of family labour, dependence of biochemical inputs, etc. (van der Ploeg, 1991). Styles of farming illustrate the variety of agricultural practices. Therefore, changes in the agricultural structure cannot be traced back to the farm without alterations (Volker, 1994).

The conclusion of this paragraph is that the effects of economic-technological and socio-political factors at the farm level are strongly transformed by differences between farmers and by variation in local physical conditions.

Farmers' reactions to the influence of the farm context

Farmers (groups) increasingly try to influence the circumstances in which they work by displaying a great variety of initiatives. Especially after 1990, farmers established more than sixty associations referred to in this chapter as *innovation groups*. These are groups of farmers aiming at a more sustainable agriculture. The name refers to the innovations the groups develop at farm level in the field of environment, nature, landscape, recreation and (regional) quality production.

Innovation groups are a reaction to the negative aspects of agriculture. The farmers involved are dissatisfied with technological and economical conditions, decreasing operational space and the way ecological problems are regulated by the government. Because of the groups' discontent and their striving towards sustainability, they can be considered as *social movements*. Social movements are defined here as relatively permanent, higher integrated and organized forms of collective action with relatively clearly articulated demands concerning social changes (Bader, 1991). The resources available to the innovation groups are manpower, internal and external financing, craftsmanship and expertise, formation of networks and regionalization.

It is not possible to give a blueprint of an innovation group. The groups differ in the specific formulation of goals, determination of tasks, approach to raising income and development of opportunities, sector, size, membership, legal status and stage of development (Migchels, 1994). Yet all these initiatives have common elements. Firstly, they are aimed at improving the quality of the environment, nature and the landscape. Secondly, they are aimed at the improvement of the farmers' income or development possibilities. Thirdly, they all aim at a new relation between authorities and farmers. Innovation groups can become important actors in a social negotiation process that has to lead to agreement on the sustainable use of land in a certain region. Besides, they fit into a new administrative culture that is no longer characterized by instrumental control ('acting on society') but by interaction between farmers and authorities (Aarts and van Woerkum, 1995).

Farmers involved in innovation groups are only a very small minority, but the innovations are very important for the development of agriculture towards sustainability. The innovative farmers are 'pioneers,' whose results are carefully watched by other farmers. Possible concrete tasks for the innovation groups are (see also Koenes, 1993; Driessen et al., 1995):

- to act as intermediate bodies between farmers and (regional) authorities and social organizations;
- to reveal and measure environmental damage at farm level;
- to adapt nature and landscape policy at farm level;
- to produce and sell regional, biological and/or quality products and develop agro-tourism;
- to advise about environmentally sound management systems, administration- and information systems;
- to stimulate research, education and information on subjects relevant to the innovation group;
- to reduce regional environmental damage by means of a tax and reward system;
- to function as a redistribution office for funds obtained from the tax and reward system and the possible enforced anti-pollution tax by the government;
- to manage and distribute emission and production rights.

Relations between innovation groups and the farm context

From 1992 to 1996 I conducted research on innovation groups in the Netherlands (Horlings, 1996). The research question of this study was: 'To what extent and in which sections does the government and its policy stimulate or restrict agricultural innovation groups in their aim for a more sustainable agriculture and how can national authorities stimulate agricultural initiatives that contribute to a more sustainable agriculture?' This question was examined by empirical research on ten innovation groups and by an evaluation of national policy directed towards more sustainable agriculture.

The innovation groups' opportunities to contribute to more sustainable agriculture are influenced by their own characteristics and by the farm context. The organizational performance of the group depends on the presence of agricultural pioneers and finance, the craftsmanship of its members, the opportunity to apply external know-how, internal cooperation between group members and the extent of regionalization of the group (Figure 8.4). Craftsmanship appears to be a prerequisite for the development of innovations in environmental, nature and landscape management. Craftsmanship is needed for quality production with less fertilizers and pesticides. The combination of care for nature and a sufficient level of production also requires specific skills. It is no coincidence that a glasshouse producer in the western part of the country won one prize for the good quality of his salad products and an environmental prize for using less water and pesticides. Quality production runs parallel to environmental innovations. However, farmers who want to implement more ecologically sound or biological production methods still face bottlenecks in management and cultivation techniques for aspects such as quality, yield and the control of weeds, disease and infestations.

	Conventional agricultural development	Innovative agriculture
Goal	raise productivity	sustainability
Resources	relatively exogenous: capital, regulation, knowledge and technology	relatively endogenous: manpower, craftsmanship and regionalization
Network	farm context	new networks and coalitions
Farmers' operational space	small	larger
Influence relations	one-sided	interaction and cooperation
Government policy	hierarchical rule	network- and self-rule

Figure 8.4: Conventional and innovative agricultural development

Innovation groups are integrated within self-created or self-selected networks. These networks give them financial means and expertise and also offer opportunities for new coalitions, for example with nature protection organizations, in order to strengthen their power *vis-à-vis* government. Since 1990 contacts with some research institutions, which supported the groups with biological knowledge or regional development plans, have been very stimulating. On the other hand, the groups have not received support from the traditional agricultural unions, in fact at least till 1995 they were rather hindered. A possible explanation for this is that the traditional unions felt threatened by the new initiatives, since the claim for more sustainable production threatened the environmental image of conventional production. Besides, when many new interest groups were arising, this would undermine their own tasks and create an unmanageable situation. Thus, innovation groups are both hindered and stimulated by actors in the farm context.

The influence of national policy on innovation groups has been evaluated in two different ways: by a personal assessment of the objectives, strategy and instruments used and by asking the innovation groups about their perception of contacts with authorities and national policy. National agricultural, nature and landscape policy changed direction from the early 1990s onwards, towards increased self-rule by the agricultural target group. Self-rule means that target groups, like farmers, take more responsibilities in environmental care. The changing environmental strategy is expressed in processes like deregulation, a more differentiated and decentralized policy and regional negotiation between authorities and citizens.

However, the set of policy measures directed towards the agricultural sector has not kept pace with this change of strategy. In particular the measures connected with mineral emissions and nature policy do not sufficiently extend the degree of *self-rule* by agricultural groups, *educational capacity* of farmers and authorities for developing an innovative sustainable agriculture and further *internalization* of environmental awareness and environmental measures.

A comparison of the results of the policy evaluation and the perception of this policy by innovation groups, shows that the groups are both stimulated and hindered as a result of government policy for rural areas. The following factors were encouraging in the early 1990s. The publication of a number of policy documents on agriculture, environment and nature around 1990 has stimulated the establishment of innovation groups. Dissatisfaction with policy appears to be a strong motive to start a group. Innovation groups communicate quite a lot with government representatives and this is generally perceived as positive. Self-rule for innovation groups opens up the possibility for influencing policy at the level of formulating objectives and concepts, and gives them some executive tasks, especially at regional level. The move from mainly judicial to a more economic and communicative type of control, supports innovative trends towards more sustainable agriculture and increases the number of behaviourial alternatives for farmers. There is increased interest in (experimental) regional policy. This creates opportunities for innovation groups to establish new coalitions and consultative structures with other interest groups within a network. A number of innovation groups receive financial support from the government authorities. Subsidies offered as part of regional policy and rural innovation, create opportunities for farmer groups as well.

The following factors relating to government policy have hindered innovation groups in the past or currently. Innovation groups are hindered by the operationalization of sustainable agriculture in national policy. National government defines sustainability as an integration of ecological and economical goals. The policy goals are aimed at a competitive and highly productive sector that diminishes environmental damage. This agricultural development model is strongly based on technological innovation and high capital inputs. This model can be regarded as a narrow operationalization of sustainable agriculture. A number of innovation groups not only emphasize income or environmental responsibility as goals, but also include other objectives in their definition of sustainable agriculture. They want to have more say in the agricultural development of their own region, stress the importance of agricultural employment, try to decrease the distance between consumers and producers or plead for solidarity with farmers in other countries. In other words, they have a broader definition of sustainability, which includes not only ecological and economic but also socio-political and cultural dimensions.

Dutch agricultural policy favours a market-oriented agriculture and more liberal international trade. This puts pressure on the price of agricultural products and is a disadvantage, especially for innovation groups which try to attach a surplus value to their regional or biological products. National policy takes insufficient notice of specific agricultural features such as the links between production and ecological processes, existing differences between farmers or the creative power of craftsmanship. The government still regards agriculture as an economic, food-producing activity while innovation groups want to broaden the concept of land-based agriculture to the management of rural areas.

The volume and rapid succession of government policy plans, rules and regional instructions are perceived by the farmers' groups as an impediment. The groups examined mentioned a number of bottlenecks, particularly related to mineral and nature protection policy. Mineral policy is regarded as involving too detailed instructions, while crop protection policy is criticized for its slowness and the high costs involved in approving new more environmentally sound herbicides. Negative views of nature policy focus on the concept of the so-called National Ecological Network, which stimulates spatial separation between agriculture and nature, and its top-down implementation. A lot of innovation groups strive for more functional integration of agriculture with nature on their farms. Their definition of nature is not limited to undisturbed large-scale reserves, but includes 'everything that lives,' not only wild plants and spontaneous nature, but also cows and grass and even the predator-insects that are used against infestations in greenhouses.

Innovation groups have also observed bottlenecks connected with the insufficient adaptation of policy areas to and integration of managerial levels. Government support to groups varies from suppression to negation. The more pragmatic groups, whose opinions and goals are closer to government policy goals, received more financial support than more critical groups. The biological groups, in particular, do not feel stimulated by the government.

Despite political changes, farmers' operational space has not been expanded. The combination of price cuts, still extensive packets of environmental measures, spatial separation of agriculture and nature, and the ever ongoing specialization and intensification of agriculture, restrict the farmers' possibilities to achieve more sustainable agriculture by self-rule (Horlings, 1996).

Innovation groups and sustainable agricultural development

Innovation groups are able to contribute to more sustainable agriculture. Some examples of innovations developed by innovation groups are a reduction in the application of minerals and chemical products for crop protection, the use of biological production methods, implementation of agricultural nature and landscape management, production of new crops or products, the development of subsidiary branches (such as farm tourism), influence on policy making and sales promotion. Innovation groups contribute to the re-linking of agricultural production methods to local conditions, for instance by adapting cultivation techniques and management to local ecological circumstances, or by developing regional quality products. Links can also be constituted between the groups' activities and organizational methods and the physical and cultural features of the region. Innovation groups desire more influence in agricultural production and processing, the sale of products and the use of local resources, such as labour and knowledge. In that sense they contribute to more endogenous agricultural development. The groups plead for self-rule over the environment, nature and landscape by taking over government tasks and responsibilities. They want to introduce care of nature and landscape to the farm, tasks that were formerly considered a self-evident part of the agricultural occupation. In this way they contribute to a 're-professionalization' of farm work. In their aim for sustainability they pursue different collective strategies, depending on their situation and tactical choices.

Because of their innovative potential and contribution to sustainable development, innovation groups deserve more support from governments. It can be expected that the call for more environmental self-regulation and responsibility in agriculture will continue. The government faces the challenge to develop a third way, between total freedom of farmers and strict regulation of production.

References

Aarts, M.N.C. and Woerkum, C.M.J. van (1995), *Communicatie en de acceptatie van natuurbeleid door agrariërs: de rol van milieu-coöperaties,* Vakgroep Voorlichtingskunde, Landbouwuniversiteit: Wageningen.

Bader, V.-M. (1991), *Collectief handelen, deel 2: sociale gelijkheid en collectief handelen,* Wolters-Noordhoff: Groningen.

Benvenuti, B. (1982), 'De technologische-administratieve taakomgeving (TATE) van landbouwbedrijven,' *Marquetalia,* No. 5.

Benvenuti, B. (1991), *Geschriften over landbouw, structuur en technologie,* Wageningse sociologische studies No. 29, Landbouwuniversiteit: Wageningen.

Dijk, G. van (1994), 'Falend beleid en "endogene" ontwikkelingen in de Europese landbouw,' *Spil,* No. 127–128.

Driessen, P.P.J. et al. (1995), *Vernieuwing van het landelijk gebied; een verkenning van strategieën voor een gebiedsgerichte aanpak,* VUGA: Den Haag.

Hayami, Y. and Ruttan, V.W. (1985), *Agricultural Development: an International Perspective* (revised and expanded version), John Hopkins: Baltimore and London.

Haan, H. de (1994), *In the Shadow of the Tree. Kinship, Property and Inheritance among Farm Families,* Het Spinhuis: Amsterdam.

Hoogh, J. de (1985/1986), 'Ontwikkelingspatronen van de landbouw,' *Cultuurtechnisch Tijdschrift,* Vol. 25, No. 4, pp. 235–243.

Horlings, I. (1996), *Duurzaam boeren met beleid; innovatiegroepen in de Nederlandse landbouw,* Circle for Rural European Studies: Wageningen.

Jongerden, J. and Ruivenkamp, G. (1996), *Patronen van verscheidenheid,* Werkgroep Technologie en Agrarische Ontwikkeling/Wetenschapswinkel: Wageningen.

Koenes, A. (1993), *De milieucoöperatie: de rol van de overheid,* Groningen.

Long, N. (1989), 'Conclusion: Theoretical Reflections on Actor, Structure and Interface,' in Long, N. (ed.), *Encounters at the Interface; A Perspective on Social Discontinuities in Rural Development,* Wageningse sociologische studies No. 27, Landbouwuniversiteit, Wageningen.

Lowe, P. and Murdoch, J. (1995), 'Networks in Rural Development,' Paper for the XVI Congress of the ESRS, Prague.

Migchels, G. (1994), 'Milieucoöperaties en natuur en landschap,' *Coöperatie,* December

Ministerie van LNV (1990), *Nationaal Natuurbeleidsplan: regeringsbeslissing,* SDU: Den Haag.

Ministerie van VROM (1989), *Nationaal Milieubeleidsplan: kiezen of verliezen,* SDU: Den Haag.

Ministerie van VROM (1990), *Nationaal Milieubeleidsplan Plus,* SDU: Den Haag.

Ministerie van VROM (1993), *Tweede Nationaal Milieubeleidsplan,* SDU: Den Haag.
RIVM (1993), *Nationale Milieuverkenning 3, 1993–2015,* Samsom H.D. Tjeenk Willink: Alphen aan den Rijn.
Ploeg, J.D. van der (1990), 'De produktie van kwaliteit: de Nederlandse pat-situatie,' in Ploeg, J.D. van der and Ettema, M. (eds), *Tussen bulk en kwaliteit,* Van Gorcum: Assen.
Ploeg, J.D. van der (1991), *Landbouw als mensenwerk; arbeid en technologie in de agrarische ontwikkeling,* Coutinho: Muiderberg.
Ploeg, J.D. van der (1992), 'The Reconstruction of Locality: Technology and Labour in Modern Agriculture,' in Marsden, T. et al. (eds), *Labour and Locality,* Fulton: London.
Roep, D., Ploeg, J.D. van der and Leeuwis, C. (1991), *Zicht op duurzaamheid en kontinuïteit; bedrijfsstijlen in de Achterhoek,* Vakgroep Agrarische Ontwikkelingssociologie, Landbouwuniversiteit: Wageningen.
Volker, C.M. (1994), 'Van halve waarheden naar een nieuwe onderzoeksagenda,' *Spil,* No. 127–128
WCED, World Commission on Environment and Development (1987), *Our Common Future,* Oxford University Press: Oxford.
Wolpert (1964), 'Decision Making in Spatial Context,' *Annals of the Association of American Geographers,* No. 54, pp. 537–538.

Part IV

Regional Development

9 Sustainable development in the Italian mountains

Giorgio Osti

Introduction

The purpose of this Chapter is to examine models and projects of development[1] in the central-northern Italian mountains, and to assess their environmental and social sustainability. Since the crisis of modernization theory, sociology has been in need of new parameters with which to understand ongoing trends of development in the peripheral areas of the more industrialized countries, especially since the emergence of the environmental question. It is necessary to understand the extent to which these models and projects of development are suited to the social and environmental conditions of peripheral areas.

This is all the more urgent in view of the shift of interest by the European Union from the agricultural sector to rural areas. Huge amounts of resources and energy are about to be committed to new projects for the development of mountain areas. The environmental question is crucial and requires clarification if new illusions, or worse, the further squandering of public money, are to be avoided.

This Chapter addresses the question of sustainable development in the Italian mountains using the *Documenti unici di programmazione* (*Obiettivo* 5b, Reg. Cee 2081/1993) of a number of Italian regions, and a survey of studies and research on socio-demographic changes in mountain areas. First I will investigate the meanings of sustainable development, then I will provide a description of the environmental and social situation of the mountain areas of central-northern Italy. This is followed by the identification of models (ideal-types) of mountain development and an assessment of the congruence between models and projects of development.

In search of the meanings of sustainable development

Sustainable development has been defined as a process which enables a population to satisfy its present needs without compromising the satisfaction of these same needs by future generations. The concept as such has not been sufficiently broken down into its individual components. One version is couched in more economic terms: the consignment to future generations of the same stock of natural capital currently available. And there is also a more ecological version: the consignment to future generations of the same variety and diffusion of species currently in existence. The former version presupposes the greater elasticity and interchangeability of goods, the latter tends towards an absolutist evaluation of the good 'nature' (Norton, 1992).

Whatever the case may be, the notion of sustainable development has restored optimism in the fate of mankind, superseding or attenuating the post-modern drift of western society (Redclift, 1993). The concept, in fact, seems to comprise two meanings. The first is reconcilability between humankind and nature. After a period during which people in the West had felt themselves profoundly and irremediably divorced from nature, the idea of sustainable development has revitalized the possibility of a balanced relationship between people and their environment. The second meaning concerns a revival of the myth of progress. This myth dominated western civilization at least since the scientific revolution. It entered a crisis both because of the environmental question (the sense of limits) and because of disillusionment with the teleological view of history (the progressive emancipation of humanity). The urgency of resolving the problems of the Third World induced revision of the catastrophist doctrine that had come to predominate in the West. The conservationist model congealed the great disequilibria between the North and South of the world.

In this cultural climate it would be foolhardy to propose a unilateral interpretation of sustainable development as a universal panacea, or as a contradictory concept exploited by the 'western powers.'[2] It is, in fact, an extremely vague concept which can be used in very disparate ways (Henke and Zappacosta, 1993). Basically, as said, it relaunches the possibility of mankind's manipulation of the environment as long as whatever is passed on to future generations is not inferior to that which is currently enjoyed. The ambiguity therefore lies in the 'value' of this heritage and in the methods used to measure it.

Economics offers methods (travel costs, hedonic, contingent analysis) which are certainly useful but nevertheless inadequate for global assessment, since the value of the environment has an imponderable dimension tied to uncertainty, as well as a relative dimension tied to tastes, cultures and states of mind. Of course, the situation cannot be left as it is – both because it, too, is the outcome of a historically determined equilibrium and because this would not satisfy the condition of development, i.e., change towards a better state of affairs.

The problem is still unresolved. The value of the environmental legacy to be left to posterity must be established case by case, area by area. Sustainable development is, therefore, an attitude, a mode of addressing reality rather than a criterion that discriminates among uses of the good 'environment.' If anything, it is a value which enhances solidarity among generations and rejects the consumerist and promethean idols of western culture. Its meaning therefore lies more in the domain of myth than in that of the analytical assessment of plans or programmes of collective interest.

The environmental and social situation of the mountain areas of central-northern Italy

Agenda 21 (UNCED, Rio, 1992), emphasizes that mountains are a particularly fragile environment, and recommends that close attention should be paid to both ecological and social questions. In this document, 'programme areas covered strengthening knowledge about ecology and sustainable development of mountain ecosystems and promoting integrated watershed development and alternative livelihood opportunities' (Yearbook of the United Nations, 1992, p. 673).

In very general terms, one may state that the mountain areas of central-northern Italy are substantially 'stable' from both an environmental and social point of view.[3] As far as the environmental situation is concerned, the great projects to exploit natural resources in the mountain areas of central-northern Italy have been concluded. The hydroelectric reservoirs have been completed; mining activity has ceased almost everywhere and is in any case uneconomical; except for the quarrying of marble, which continues in some circumscribed areas. This has created some problems of compatibility, especially as regards protected areas and ones of environmental emergency. The most serious threats are raised by infrastructural projects for transport (roads and railways) and by the ever greater presence of tourism. The quiet holiday in the mountains has now given way to multiple uses of the environment, and in every period of the year. The use of the 'mountain bike' is an emblematic example, since it is a sport which is in some respects ecological but which nevertheless occupies more and more natural space. In addition to mountain biking there is the use of horses, motorbikes and off-road vehicles, as well as off-piste skiing, helicopter skiing, free climbing, and so on. It is difficult to assess the environmental impact of tourism and the ability of ecosystems to sustain it. Conversely, it is rather easy to establish the psychological threshold of tolerance of overcrowding, which is greatly exceeded in many localities at certain times of the year.

Tourism has conquered every area of the mountains. Indeed, it is not difficult to find tourists in even the most inaccessible places. Nevertheless, the frequency of use is highly variable. Overcrowded areas adjoin others in which the human presence is almost negligible, largely because of emigration over several decades and the resettlement of residents in the valley bottoms, although this is not the case with certain ski resorts, where the phenomenon has occurred in reverse.

However, the descent to the valley bottoms of the population and the abandonment of agriculture at higher altitudes has led to the re-naturalization of many mountain areas, especially in the Apennines, where winter sports are much less widespread than in the Alps.

This abandonment of permanent high-altitude activities (agriculture, cattle raising and residence) has encouraged the growth of woodland and the return of animals – like wolves, bears, eagles and lynxes – which had long been extinct. Wolves have now re-established themselves along the Apennine chain as far as the French Alps. Bears and lynxes have been reported in the eastern Alps close to the border with Slovenia and Austria. These are clear indicators of the growing naturalness (completion of the food chain) of the mountain environment, although it is for the moment difficult to determine the extent to which the presence of these predators will be permanent.

In conclusion, the human population has partially withdrawn from the mountains, visiting them with increasing frequency at the weekends and for holidays. Tourism brings considerable risks but it is also rather elastic, in the sense that it is can be regulated (e.g., by restricting the number of visitors), redirected (e.g., towards 'softer' tourism) and educated (e.g., with courses in environmental awareness), thereby greatly reducing its impact. Large-scale construction projects or quarrying, however, meet with growing opposition – from local populations as well, with their increasing reluctance to accept monetary compensation or short-term economic development. The mountain areas of central-northern Italy, therefore, exhibit a not particularly negative environmental balance, although the equilibrium is still extremely precarious. Constant monitoring and containment of the micro factors of natural and human degradation is necessary. A looming threat in most valleys is hydro-geological instability, which is enormously costly to deal with. At times, remedies introduced in one location increase the risk of landslides in another. The most evident example is the cementifying of watercourses: erosion is reduced at the points where the banks have been made watertight, but the risks of flooding and land slippage increase downstream.

One also notes relative stability at the social level. The great exodus from the mountains has come to an end. Indeed, in almost all mountain communities the migratory process now operates in reverse: more people are settling in these communities than leaving them. Although lower than the average for the region, the level of affluence is nevertheless such to guarantee a decent lifestyle, and the standard of housing has risen enormously (Barberis, 1992). The level of unemployment is not particularly high, or at least no higher than that of the surrounding regions as a whole. Community services are satisfactory although a growing need for rationalization has led to a certain amount of centralization, which means that travelling times have considerably increased for a low density population. This has had especially negative repercussions for the less mobile groups in the population (old people and children).

The most serious social problems in the central-northern mountain areas are the ageing of the population and the existence of extremely small villages inhabited solely by elderly people. The migratory inflow of recent years has not improved this situation, since, although more people are arriving than leaving, the former are older than the latter. Moreover, the new arrivals tend to settle in relatively more accessible mountain areas. With the passage of time, the imbalance between age classes gives rise to two problems: one economic-organizational in nature (the costs of satisfying the needs of an elderly population), the other more specifically cultural (self-representation in such an imbalanced society, without heirs).

Development and sustainability in mountain areas

The sustainability of mountain areas depends on the extent to which two problems – the demographic imbalance of the population and the pervasiveness of tourism – can be kept under control, without affecting local incomes. This statement rests on a set of cultural values embraced by mountain people, which can be regarded as postulates for sustainability:

- the population's right to live in the mountains and to enjoy a standard of living similar to that of urban areas;
- closer control over a long occupied territory by means of a stable human presence;
- counteracting 'grab-and-run' mass tourism by a stable human presence that provides further guarantees of environmental protection.

		Territorial scale of the factors	
		Local	Overlocal
Causes of development	Factors linked to residence	*Neo-ruralism* Making the most of rural culture and communitarian way of life	*Metropolitan development* Lower costs of housing; good roads to centres; search for good quality environment
	Factors linked to work	*Endogenous development* Making the most of local resources; keeping within the area the wealth; developing local cooperation	*Productive decentralization* Less costly and more compliant workers; search for public funds

Figure 9.1: Typology of development models in Italian mountain areas

Environmental sustainability is therefore tied to social sustainability, that is, to the possibility that a stable community can live in the mountains with dignity. What is meant by a 'stable community' is by no means clear: the number of inhabitants, the form of their interaction, and the means or level of their sustenance are not pre-ordained.

However, difficult it is to establish precisely the point of arrival of social development ('a stable community'), it is at least possible to assess its progress. One may usefully investigate the factors responsible for the greater equilibrium between mountain areas and lowland areas as compared with the past. The exodus from most of the central-northern mountain areas has now ceased. Indeed, the population inflow was already greater than the outflow by the 1980s. Areas which for decades were affected only by emigration are now faced by a movement in the opposite direction.

The overall demographic ratio is still negative, however, because the birth rate is very low while the share of the elderly population is very high. Small isolated communities continue to be affected by a negative migratory ratio. Nonetheless there is no doubt that the trend has reversed, and that it is matched by a pronounced process of delocalization[4] in the whole of central-northern Italy. What are the causes of this phenōmenon as regards mountain areas in particular?

Figure 9.1 sets out two causative factors of peripheral development; one linked to residence, the other linked to work. The impetus for finding a residence and an environment more congruent with one's ideals and one's principles of environmental quality, concerns not only the comfort of the home but also quality of the natural and social environment in which it is located. The impetus deriving from the search for income, professionalism and satisfaction that derive from work is also not merely a matter of instrumental motivation but concerns work in its entirety (pay, work environment, intrinsic satisfaction).

These two factors cross-reference at both local and supralocal levels on the territorial dimension. This is evidently an enormous simplification in an age of worldwide integration. Nevertheless, one may use 'local' to refer to the mountain valley, as the geographically homogeneous district that groups a number of mountain villages around a small urban centre, also located in the mountains. A district of this kind often has a historical and administrative identity (the 'mountain community') too. The supralocal level refers to a much broader area comprising uplands, lowlands, urban centres and even metropoles.

The four quadrants thus created contain four models (ideal-types) of the development of peripheral areas. The first is the neo-rural model, where the progress of an area is driven by the presence of people with strongly cultural motivations, who embrace the ideals of rurality (communitarian life, simplicity and sobriety, spirituality, and so on). Work also belongs to this ideal, but it is part of a larger whole that must integrate harmoniously with the life rhythms of nature and of the local community (Leger and Hervieu, 1980; Amadei, Corazziari and Montanari, 1981; Chevalier, 1981; Mercier and Simona, 1983; Blackwood and Carpenter, 1978). Lack of space precludes further specification of the real or presumed components of this rurality here. There is no doubt, however, that there has been a movement in this direction, starting in the 1970s and continuing in different forms and with different motivations,[5] however small it may be in quantitative terms.

The second model, extremely powerful ideal in mountain areas, is that of endogenous development which embodies the goal of political autonomy and the enhancement of the economic and entrepreneurial resources of the residents. Movements for the redemption of the Third World and peripheral areas have been powerfully inspired this model, to the extent that it is now part of the official ideology of development organizations. The more recent concern for the natural environment, as a local resource that should be appropriately exploited by local residents, also derives from this model. The emphasis is on the production of wealth and on the local administration and enjoyment of this wealth. The ideal-typical features of endogenous development comprise the presence of numerous small firms, the local control of accumulation, and the vertical integration of the production-valorization of local goods.[6]

The third model shifts attention to the relationship between the centre and the periphery. The relative stability of mountain areas is made to depend on the metropole (Frisbie and Kasarda, 1988). Cities have expelled residents, who have occupied areas that were once marginal and this has occurred on diverse territorial scales, because the magnitude and dynamism of urban poles is different (Osti, 1995). It is not just that new residents have arrived from the cities, but the inhabitants of mountain areas themselves find it more convenient to travel distances to work while continuing to live in the mountains. The environmental advantages of this choice are indubitable: house prices are lower, and certain comforts once only offered by urban housing (running water, electricity, accessibility) are now widely available in even the remotest mountain villages. Enormous effort has gone into building farm roads. Those who move to or remain in the mountains are willing to put up with the discomforts of commuting in exchange for certain environmental advantages. It is thus extra-work conditions, as well as lower crime levels, and the chance of living on a smaller and more human scale, which attract the ex-metropolitan newcomers. One should also bear in mind the extremely large numbers of second homes in many mountain areas, which effectively become first homes during certain periods of the family life cycle (Dallari and Gaddoni, 1994). The most salient phenomenon, however, is the arrival of retired people who settle in mountain areas for affective reasons (place of origin) or environmental ones (healthfulness).

Finally, there is the model of productive decentralization, which was an Italian phenomenon typical of the 1970s (Del Monte and Raffa, 1977). Firms have located productive units in even the remotest areas due to trade union inflexibility or the generous incentives for industrialization offered by the public authorities. It is no longer necessary for the inhabitants of mountain areas to change residence in order to find work, which is available locally or, more frequently, just beyond the valley. Opinions about this type of industrialization differ: there are those who emphasize its dependence and precariousness, while others view it as a stimulus that has boosted local initiatives, and without which nothing would have happened in many peripheral areas (Gubert, 1994). In some cases, large high-tech firms have located production units in peripheral areas, thereby radically transforming the local economy. In other cases, however, large firms from outside have neither stimulated the energies of local businesses nor activated subcontracting processes. Much clearly depends on location and on the control of such rich resources as, for example, management and research (Lowe and Murdoch, 1995); and much also depends on the ability of the local economy to structure itself into a system or district (Bagnasco, 1988; Conti and Julien, 1991).

The model of metropolitan development is the one that interprets the most established trends in the mountain areas of central-northern Italy. The most important factors inducing large numbers of people to remain in, or move to, mountain areas close to urban poles/networks are better communications, the crisis in the ideology of the city, the desire to return to one's place of origin on retirement, the growing importance of environmental factors, and good pensions during the 1980s.

Among the numerous indices of stability are the positive migratory ratio, together with the fact that the mountain productive system, starting with agriculture, is stagnant. The ratio between the numbers of jobs created and lost in the last inter-census period is, at its most favourable, in equilibrium. Large numbers of residents commute to work, although not solely in the city centres. Thanks to the dispersion of productive activity, many jobs have become available on the fringes of mountain areas (Dematteis, 1988).

In short, the mountain areas of central-northern Italy have stabilized: they have not experienced economic growth but nor have they lost population. The migratory outflow has almost halted, while the inflow consists mainly of people in search of a new residence. Of course, there are dozens of local factors of development: the cooperative movement, local banks, traditional crafts, good schools, communications, public investments, and so on. These, however, are typical only of each single area. It is the suburbanization of the mountains with respect to their surrounding upland belt of diffused industrialization, to urban and metropolitan centres, and to areas highly specialized in tourism and agriculture that seem to be the most influential (Dematteis, 1988).

Against this background, we may now consider whether these trends meet the requirements of sustainability. If the residential model effectively predominates, the environmental sustainability of mountain areas becomes more certain.[7] If the choice of living in mountain areas is dictated by their environmental qualities, then the endeavour to maintain this heritage intact will be greater. If there is no pressure to establish or expand productive activities, management of the environment becomes more straightforward; indeed, nature becomes the strategic resource of central importance to residents. In this case, agriculture is conceived and valued as the custodian of tradition and of the landscape. There is little tolerance for tourism, even in its 'softest' forms, because it invariably involves invasion of an environment felt to be one's own. It should be pointed, however, that there are large numbers of semi-residential tourists; that is, people who spend most weekends at their homes in the mountains.

The models of development based on work obviously have a greater environmental impact. Theoretically, productive decentralization should create greater risks for the environment because instrumental motivation (that is, exploiting the environment in order to produce at lower costs) that predominates.[8]

Projects for development and their sustainability

Having analyzed the main trends of development in mountain areas, and briefly considered their hypothetical sustainability, we shall now examine some development projects. The question is whether projects for the development of mountain areas drawn up by local authorities take account of development trends. We can assess the congruence between trends and projects of development, and the many ways in which the two combine by imagining the following three situations:

A the development trend has high sustainability and intervention increases it;

B the development trend has ambivalent consequences for sustainability, and intervention is partial;

C the development trend has low sustainability, and intervention seeks to act against this.

EU development projects devote a great deal of space to sustainable development. It is posited as a goal and as a criterion with which the various projects must comply. For example, under objective 5b (development and structural adjustment of rural areas) plans for development must contain both an assessment of the environmental situation and the envisaged impact of the strategy or action. Moreover, these plans must indicate the measures taken by member-countries to involve their authorities responsible for environmental matters.

Examination of the projects drawn up by a number of Italian regions[9] for objective 5b (years 1994–99) shows that sustainable development is constantly cited but never explained (the EU speaks of principles of sustainable development). The expression is probably regarded as self-evident, and it is generally used as synonymous with a concern for the environmental impact of the intervention. This reinforces the point made at the outset that the appeal of the expression 'sustainable development' lies in its coupling of two terms long considered irreconcilable but nevertheless unavoidable by the public planner.

Assessing the orientation of the EU and the Italian regions according to the trends for development of mountain areas produces different results. There is an apparent mismatch between the projects proposed and what one observes in reality. The philosophy of regional plans behind 5b largely correlates with the 'endogenous development' quadrant of Figure 9.1. (Molla, 1993, p. 59). It is mentioned on various occasions in the plans, but it is most evident in the values

Table 9.1: Spending forecasts of 5b objective projects according to region and sub-programme (values in 1,000 Ecus)

	Region				
Sub-programme	Piedmont	Liguria	Emilia-R	Tuscany	Veneto
Agriculture	79.77	23.576	83.355	167.778	273.3
Rural space, forests and grazing	–	17.029	–	36.175	–
Environment	53.36	9.986	17.456	58.000	28.0
Industrial activities	96.08	105.577	127.777	241.130	379.7
Tourism	58.23	22.105	64.785	169.500	105.9
Cultural goods	–	–	–	27.900	–
Infrastructures for firms	19.91	–	–	–	36.9
Human resources	21.23	9.127	16.117	40.444	46.3
Assistance and monitoring	2.93	2.158	2.100	4.000	2.7
Total	331.51	189.558	311.590	744.927	872.8

assigned to the various areas of intervention (Table 9.1.). The area of intervention – or as 5b puts it, the 'sub-programme' – to receive the greatest amount of investment is support for non-agricultural and non-tourist activities (industrial activities or manufacturing). In the five regions examined, the amounts assigned to this sector are never less than 30 per cent of the total expenditure envisaged (Figure 9.2), followed by 25 per cent for agriculture (including forestry and

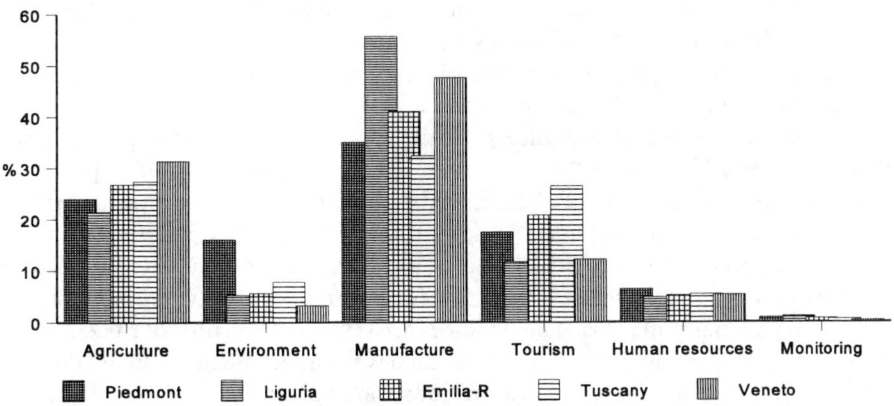

Figure 9.2: Spending forecasts of EU 5b objectives for sub-programme and region

cattle raising). Third place is occupied by subsidies to tourism, while around 5 per cent is allotted to the environment and to training. However, if one examines expenditure for environmental action across the various sub-programmes (generically called 'axis' spending), one notes considerable increases and sizeable differences among regions. Piedmont earmarks 22 per cent for the environment but Veneto little more than 7 per cent.

The new generation of development projects is therefore clearly oriented towards the diversification of economic activity. Agriculture has lost its traditional primacy to second-sector activities. There is considerable emphasis on increasing local products and on environmental protection, although not all regions make a corresponding financial commitment. In short, one may speak of projects that aim to reach endogenous, integrated and eco-sustainable economic development.

The mismatch arises from the fact that spontaneous development of the mountain areas of central-northern Italy apparently stems from other factors, which have little to do with the dynamics of private industry and much more to do with life conditions and generous pensions. So trends are nearer factor 'residence,' while projects are nearer factor 'work.'[10]

This dissonance between the projects and the actual situation of mountain areas prompts two contrasting observations. The first is that, given the fragility of development based on residence alone, it is advisable to integrate such development with subsidies to industry. If the factor 'residence' is to hold, the factor 'work' must be boosted. Spontaneous development must be backed by development guided by public planning. The opposing observation is that support for industry gives rise to dependence, because the firms thus created do not have a real market in which to compete, and end up depending on public aid. The massive subsidization of small firms in mountain areas is self-defeating, with high rates of bankruptcy, failure and outside exploitation.

It is not easy to choose between these two interpretations without a great deal of detailed information concerning the support provided to firms in disadvantaged areas. What is certain, however, is that the projects' philosophy proposed by the EU and more or less faithfully adopted by certain central-northern Italian regions is not incorrect, although it is perhaps incomplete. It rightly focuses on production, the protection of the environment, and the promotion of tourism, but it neglects the factors of development that derive from increased residence in mountain areas. These factors are tied to the quality of the environment – and in this case reality matches the development plans – but also to the quality and cost of housing, communications, the availability of certain effectively and symbolically crucial services (such as primary schools, and post offices).

More generous housing subsidies should be offered.[11] Recent trends indicate that it is easier to attract residents to mountain areas than firms, probably because less onerous incentives are required. This policy would also deal with the problems of depopulation and imbalance among age classes. The arrival of new residents would enliven the small mountain communities. Targetting residence rather than on work would obviate a further problem: many measures to stimulate the economy in mountain areas lack one essential ingredient: entrepreneurs.

It might be objected that it is not the purpose of European Union programmes, and of objective 5b in particular, to remedy the shortcomings of local services. This is the responsibility of the authorities concerned. Although this is true, the Italian *Comunità Montane* (public authorities comprising homogeneous mountain communes) took action on these matters in the 1980s by defending the environment and developing civil infrastructures (Osti, 1996). However, the resources with which to pursue this work are by now few, and there is a tendency to curtail services in sparely populated areas by closing numerous shops, schools, small hospitals and post offices.

The problem, perhaps, is to understand which notion of rural development is entertained by the promoters of EU policies. One suspects that they still give priority to a 'productivist' stance that encompasses environmental awareness but fails to consider the other social and cultural needs of a community and their interaction with the economic sphere.

We may conclude that what we have here is probably case 'B': the development trend has ambivalent consequences on sustainability and the intervention is partial. The suburbanization of mountain valleys moves the border line between marginal and developed areas. This line used to be between mountain and plain, now it is within the mountains. Polarization within the mountain areas, according to the postulate 'presence of people maintains the environment,' reduced sustainability in the most depopulated ones. Development policies, particularly 5b projects, take account only of the economic aspects of these areas.

Conclusions

Some final remarks are in order concerning the priorities set out in the EU plans for development. The prevalence of economic considerations, albeit intersectoral and eco-sustainable, is probably due to two factors. The first is that these plans are formulated within an agricultural-economics tradition. Public agencies, offices, personnel, even at the level of the European Union itself, have an agricultural rather than rural mentality. That mentality arose during the period when agriculture was being modernized, and the greatest effort was directed towards introducing innovations to increase production. This initial pattern may then have evolved into current models of development – diversification being a case in point – but the underlying logic is still markedly economic.

The second factor concerns the ability of the local political apparatus to interpret changes in the mountain environment and to introduce measures appropriate to them. Attention has rarely been paid to the new residential function of mountain areas. It has been easier to focus on the ideal of endogenous development: more autonomy, more localist pride, greater pressure on the central government. This process was extremely useful during the period when mountain areas, where significantly marginalizes, since it enabled the mountain populations to acquire dignity. By now, however, endogenous development is apparently not enough. Mountain areas also show pronounced signs of integration with central ones, entailing a change of strategy about which local elites are probably unaware.

Analysis of discrepancies between models and projects of development in rural areas can only be highly generic. It is perhaps foolhardy to seek to identify general trends in the Italian mountain areas when their development is still tied to specific local factors or to supranational forces (globalization). It is also certainly too early to assess the effective outcome of the regional programmes implemented by the European Union, since much depends on how this implementation comes about and on the actors involved in the process.

Nevertheless, one may usefully reflect on 'philosophies' of development and on their congruence with certain demographic trends. If, as this Chapter has suggested, the sustainable development of mountain areas requires the stable presence of people and the main trend is 'metropolitan development' (urbanization of most accessible mountains), then the most urgently needed measures are those designed to attract young people to settle in further areas. This entails – especially if these young people have small children – the maintenance of services without which even people willing to commute over long distances to work would decide not to make the move. Forms of mixed incentive (housing + work) must be devised, coupled with the promotion of public services run by private non-profit organizations (Salais, 1993; Jäger 1995; Federazione Trentina delle Cooperative, 1996). There are numerous initiatives available to encourage the spontaneous tendency to return and settle in the Italian mountains. No development project that claims to be 'sustainable in the fullest sense' can, however, ignore the social dimension of living in the mountains.

Acknowledgement

Research funded with the support of the National Research Council of Italy, Special Project RAISA, Subproject no. 1, paper no. 2746.

Notes

1. The word 'development' is intrinsically ambivalent because it embodies two aspects: description of a change in one direction, making it an analytical tool; description of a change towards a desirable situation, giving it a normative or ideal content. In this Chapter the term 'model of development' refers to an analytical tool that helps describe what happens (trends) in the Italian mountains, while, the term 'project of development' refers to actions that seek to shift development towards a desirable situation.
2. The powerful egalitarian thrust (environmental justice) apparently inherent in the concept of sustainable development should forestall accusations of cultural imperialism by the West (Bruntland, 1989). However, several writers have stressed that assessments of sustainability are hazardous unless they take account of cultural differences when present and future needs are defined (Meredith, 1992; Redclift, 1994).
3. 'Stable' means that central-northern mountain areas are not undergoing rapid and dramatic changes. The mountain areas of southern Italy have a very different situation. The population is still migrating from the mountains and concentrating in urban areas and along the coasts (Celant, 1992).
4. According to Frey (1987), the delocalization process can be interpreted from two perspectives, which he calls 'regional restructuring' ('polarization of growth tendencies between large command and control center areas and subordinate centers' p. 242), and 'deconcentration' ('new production space will be shaped much more than in the past by the residential location preferences of workers and resident-consumers' p. 243).
5. An example is provided by the numerous foreigners, mainly German and English, living more or less permanently in remote villages in the Tuscan and Ligurian mountains. The inflow into the mountains of young people seeking to set up farm businesses, or today tourism-environmentalist activities, still continues.
6. See Garofoli, 1992. Examples in Italian mountain areas are spectacle manufacture in the Cadore, tourism in the Dolomite area, fruit growing in Trentino, the production of parmesan cheese in Emilia, and industry in the hinterland of the Marches (Scaramellini, 1995). Special examples are provided by cow raising (Ventura, 1995) and green tourism (Van der Straaten, 1992).
7. With a difference, however, between the neo-rural and metropolitan models. The latter tends to produce a dualism between mountain areas more accessible to services and work and those less so.
8. The model of autogenous development may be the most sustainable inasmuch as production and consumption tend to overlap in the same area. In this case, energy and transport costs are lower and maintaining equilibrium between natural resources

and consumption is greatly in the interest of the local community. However, this form of autarky theorized by human ecologists (Berkes, 1989; Meredith, 1992) is difficult to achieve in marginal areas surrounded by areas of strong development.

9. The following documents were consulted: *Regione Liguria, Documento unico di programmazione obiettivo* 5b, 13 aprile 1995; *Regione Piemonte, Documento unico di programmazione 1994–99*, Torino aprile 1994; *Suppl. Straordinario BUR regione Toscana* n. 10, 1.2.1995; *Documento unico di programmazione Suppl. BUR regione Veneto* n. 42, 29 giugno 1994; *Documento unico di programmazione regione Emilia-Romagna*, aprile 1994.
10. In its *Documento unico di programmazione* for 5b objective, the region of Tuscany demonstrates a certain awareness of the need not only to promote the economic system but also to improve living conditions in mountain areas. It states, relative to the priority axis of development: 'strengthening of extant or potential industry as well as social and environmental aspects by taking concrete action to remedy the inadequacy and on occasion lack of those services indispensable for good living and working conditions.' (*Suppl. Straordinario BUR regione Toscana* no. 10 of 1.2.1995, p. 44). The Region of Emilia Romagna also confronts the problem: 'The plan uses all the instruments and opportunities made available by reform of the structural funds, both those to support industry and those intended to improve housing conditions and services' (*Documento unico di programmazione, regione Emilia-Romagna*, April 1994, p. 20). However examination of the practical measures envisaged by both regions reveals that none of them concerns housing and services.
11. As provided for by the recent Italian law on mountain areas (*Legge Nazionale* no. 97/1994), although it has proved difficult to implement mainly in the sector of housing incentives.

References

Amadei, G., Corazziari, G. and Montanari, A. (1981), *Cooperative di produzione e agricoltura di gruppo*, Angeli: Milano.

Bagnasco, A. (1988), *La costruzione sociale del mercato. Studi sullo sviluppo di piccola impresa in Italia*, Il Mulino: Bologna.

Barberis, C. (1992), *Montagna 2000*, Angeli: Milano.

Berkes, F. (ed.) (1989), 'Cooperation from the Perspective of Human Ecology,' in Berkes, F. (ed.), *Common Property Resources. Ecology and Community-Based Sustainable Development*, Belhaven: London.

Blackwood, L.G. and Carpenter, E.H. (1978), 'The Importance of Antiurbanism in Determining Residential Preferences and Migration Patterns,' *Rural Sociology*, Vol. 43, No. 1, pp. 31–47.

Bruntland, G.H. (1989), 'Sustainable Development: An Overview,' *Development*, No. 2–3, pp. 13–14.

Celant, A. et al. (1992), 'Caratteri generali e dinamica recente del fenomeno urbano in Italia,' in Dematteis, G. (ed.), *Il fenomeno urbano in Italia: interpretazioni, prospettive, politiche*, Angeli: Milano.

Chevalier, M. (1981), 'Le phénomènes néo-ruraux,' *L'Espace Géographique*, No. 1, pp. 33–47.

Conti, S. and Julien, P.A. (eds) (1991), *Miti e realtà del modello italiano: letture sull'economia periferica*, Patron: Bologna.

Dallari, F. and Gaddoni, S. (1994), 'Residenze ed insediamento tra recupero e nuova presenza nell'Appennino emiliano-romagnolo,' in Bernardi, R., Salgaro, S. and Smiraglia, C. (eds), *L'evoluzione della montagna italiana fra tradizione e modernità*, Patron: Bologna.

Del Monte, A. and Raffa, M. (1977), *Tecnologia e decentramento produttivo. L'esperienza italiana dal dopoguerra ad oggi*, Rosenberg & Sellier: Torino.

Dematteis, G. (1988), 'Valorizzazione e trasformazioni territoriali,' in Leone, U. (ed.), *Valorizzazione e sviluppo territoriale in Italia*, Angeli: Milano.

Federazione Trentina delle Cooperative (1996), *Analisi sul sistema distributivo cooperativo in località marginali*, Trento.

Frey, W.H. (1987), 'Migration and Depopulation of the Metropolis: Regional Restructuring or Rural Renaissance?' *American Sociological Review*, Vol. 52, No. 2, pp. 240–257.

Frisbie, W.P. and Kasarda, J.D. (1988), 'Spatial Processes,' in Smelser, N.J. (ed.), *Handbook of Sociology*, Sage: London.

Henke, R. and Zappacosta, M. (1993), 'Crisi della modernizzazione e sostenibilità nel terzo mondo: alcune riflessioni sul settore agricolo,' *La Questione Agraria*, No. 51, pp. 17–44.

Garofoli, G. (1992), *Economia del territorio. Trasformazioni economiche e sviluppo regionale*, Etas: Milano.

Gubert, R. (1994), 'Problemi e politiche dello sviluppo rurale: aspetti sociologici,' XXI Convegno Sidea, Campobasso 22–23 settembre.

Jäger, H. (1995), 'Il futuro dei servizi in ambito rurale,' *Leader Magazine*, No. 9, pp. 6–12.

Leger, D. and Hervieu, B. (1980), 'Recours à la ruralité et crise,' *Economie Rurale*, No. 140, pp. 16–20.

Lowe, P. and Murdoch, J. (1995), 'Networks in Rural Development,' paper XVI Congress of the European Society of Rural Sociology, Prague, 31 July–4 August.

Mercier, C. and Simona, G. (1983), 'Le neo-ruralisme. Nouvelles approches pour un phenomene nouveau,' *Revue de Geographie Alpine*, No. 3, pp. 253–265.

Meredith, T.C. (1992), 'Environmental Impact Assessment, Cultural Diversity, and Sustainable Rural Development,' *Environmental Impact Assessment Review*, No. 12, pp. 125–138.

Molla, L. (1993), 'Riforma dei Fondi strutturali Cee e implicazioni per la valutazione della spesa regionale,' in Florio, M. and Robotti, L. (eds), *Valutazione della spesa regionale e riforma dei fondi strutturali Cee*, Il Mulino: Bologna.

Norton, B. (1992), 'Sustainability, Human Welfare, and Ecosystem Health,' *Environmental Values*, Vol. 1, pp. 97–111.

Osti, G. (1995), 'Le metropoli italiane viste dalla montagna,' *Sociologia Urbana e Rurale*, No. 47, pp. 121–140.

Osti, G. (1996), 'L'azione delle comunità montane,' *Amministrare*, No. 1, pp. 123–142

Redclift, M. (1993), 'Sustainable Development: Needs, Values, Rights,' *Environmental Values*, Vol. 2, pp. 3–20.

Redclift, M. (1994), 'Sustainable Development: Economics and the Environment,' in Redclift, M. and Sage, C. (eds), *Strategies for Sustainable Development. Local Agenda for the Southern Hemisphere*, Wiley: Chichester.

Salais, J. (1993), 'Economie des services publics dans l'espace rural,' *Futuribles*, September, pp. 45–49.

Scaramellini, G. (1995), 'Concluding Remarks,' in Scaramellini, G. (ed.), *Sustainable Development of Mountain Communities*, Guerini e Associati: Milano.

Van der Straaten, J. (1992), 'Appropriate Tourism in Mountain Areas,' in Briassoulis, H. and Van der Straaten, J. (eds), *Tourism and the Environment*, Kluwer: Dordrecht.

Ventura, F. (1995), 'Styles of Beef Cattle Breeding and Resource-Use Efficiency in Umbria,' in Van der Ploeg, J.D. and Van Dijk, G. (eds), *Beyond Modernization*, Van Gorcum: Assen.

Yearbook of United Nations, (1992), New York.

10 Rural tourism and sustainable development: a case study on Slovenia

Alenka Verbole

Introduction

In recent years there has been a renewed interest in developing the rural areas of Europe. The reasons for this concern lie in the profound changes that are affecting the European countryside and rural societies. In Europe, migration to the cities has eroded the vitality of rural communities. Traditional methods of farming are falling into disuse, the quality of the environment is deteriorating, and income and employment opportunities in rural communities are decreasing.

Many European countries began to look for alternative and (it was hoped) more profitable economic activities that would help to revitalize the countryside and rural communities in the mid-1980s, as it became obvious that the agricultural sector alone did not hold the key to rural development. One of the main strategies of this search was to identify ways of encouraging the diversification of rural economic activities. Rural economic diversification brought with it an interest in tourism as a tool to revitalize the countryside and rural communities.

In Eastern and Central European countries especially, rural economic diversification was complicated by the problems resulting from political, economic, and social transformations. The recent governmental efforts to switch from a centrally planned to a market-oriented economy (see Barbič, 1993, 1992), the growing trend toward privatization and decentralization, and the cutbacks in social benefits[1] have strongly influenced and changed the basis of economic and social life in these societies.

The shift in governmental policy was speeded up by the growing numbers of city people who wanted to spend their holidays in the country.[2] Various forms of rural tourism – agri-tourism (in which suppliers are directly engaged in farming), farm tourism (in which suppliers are farm owners), and village tourism (in which the entire village is the supplier)[3] – began to flourish in rural areas. As a result, the traditional role of rural communities, the rural environment, and agriculture changed. In fact, they are still changing. They are a resource for, and a potential beneficiary of, tourism. Today, the rural environment and farming communities are being used not only to meet the food needs of the population, but also to meet the demands of tourism, including recreation and leisure activities (Cloke and Goodwin, 1992). The rural scenery – a mix of agriculturally patterned landscapes, forests, open spaces, and picturesque villages – and the human and cultural capital of the local communities are the two main ingredients of the 'rural idyll' that attracts tourists to rural areas.

As rural tourism develops, it breaks down the stereotypes that exist about the countryside (Dernoi, 1983) and rural societies (Jansen-Verbeke, 1990, 1991). The dynamic development of rural tourism is changing the economic, social, and cultural make-up of rural societies and affecting their natural settings. These changes, and their impacts, can be either positive or negative. On the positive side, they include new opportunities for local people to increase their income (Vafiadis, 1992, quoted in Giagou, 1996), the preservation of the landscape, and the fostering of traditional activities to be 'gazed upon' (Urry, 1990). On the negative side, they range from the complete cessation of existing agricultural activities (te Kloeze, 1994) and the inflation of land prices (Dernoi, 1983; Jansen-Verbeke, 1990) to the shift in labour away from agriculture and towards the service industry (Murphy, 1985), leading to the commercialization of rural life (Pearce, 1990). Before we begin a closer examination of the costs and benefits of tourism development in the countryside, let us consider the problem of defining what rural tourism really is.

Principal forms of rural tourism

What is rural tourism? At first glance, the answer to this question is simple: rural tourism is tourism which takes place in the countryside (Lane, 1994). But the reality is more complex. Rural tourism has many characteristics, the main ones being that it is developed in non-urban areas and it is based on the exploitation of rural resources, namely, nature, local people, and their culture. Because of this variety of characteristics, Keane and Quinn (1990) were right to observe that rural tourism does not yet have a distinct image. In the literature there are many terms to describe touristic activities in rural areas (Keane and Quinn, 1990; Keane, 1992): 'agri-tourism,' 'farm tourism,' and 'countryside tourism,' to

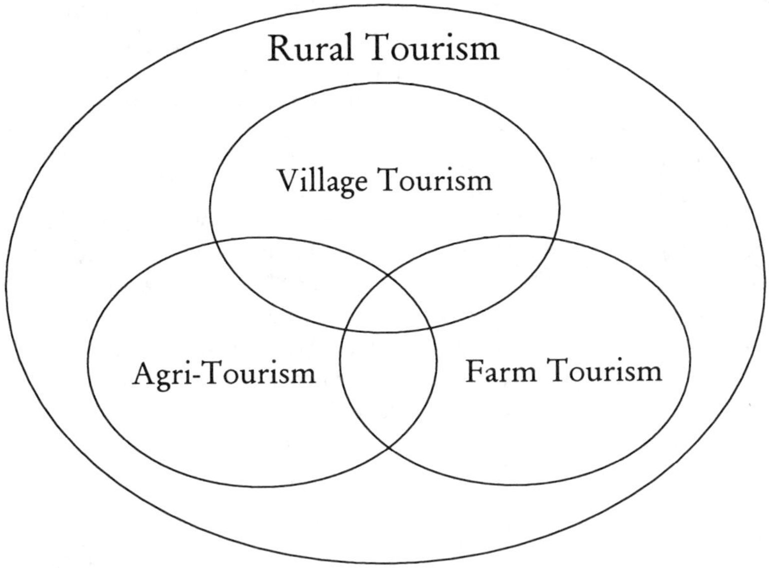

Figure 10.1: The interactions of different forms of rural tourism (after Bubendorf, 1993)

mention only a few. Moreover, the mentioned terms seem to have different meanings in different countries (Lane, 1994; Bramwell, 1994). Despite suggestions that agri-tourism was the best market description of tourism in rural areas, none of these terms has won acceptance in all parts of Europe (Lane, 1994).

For the purposes of this Chapter, and to avoid confusion, we shall adopt the definitions formulated by Jansen-Verbeke (1991).

- *Rural tourism*: all forms of tourism that are found in the rural area;
- *Agri-tourism*: all forms of tourism that are directly connected to the agrarian environment and agrarian products. The suppliers of this tourist product are directly engaged in farming;
- *Farm tourism*: all forms of tourism that are directly connected to a farm (e.g., meals at the farm, entertainment at the farm, day trips to the farm).

To these concepts we shall add that of 'village tourism' (Kovacs, 1993), which is very popular in the Hungarian countryside, and which falls into the category of rural tourism as well. Figure 10.1 shows how the different forms of rural tourism interact.

Impacts of rural tourism: pros and cons of rural tourism development

> Lord Jesus, Son of God, have mercy on the cities, the islands, and the villages of this Orthodox Fatherland, as well as on the holy monasteries which are scourged by the worldly tourist wave. (Greek Orthodox Church prayer recommended in the 1970s quoted in Urry, 1992).

Mathieson and Wall (1982) group the impacts of rural tourism into three major categories, namely economic, environmental, and socio-cultural. Economic impacts are associated with the monetary costs and benefits that result from the development and use of tourist facilities and services. Environmental impacts are associated with the ways in which tourism contributes to alterations in both the natural and the man-made environment. Socio-cultural impacts are usually associated with the ways in which tourism contributes to the changes in value systems, collective behaviour, family relationships, collective lifestyles, safety levels, moral conduct, creative expression, traditional ceremonies, and community structure.

In reality the differences between these categories may be somewhat artificial. At times, the boundaries between the categories can blur and their contents can overlap. An influx of money into the local economy, for example, can be classified as an economic impact *and* a social impact, as the availability of more money, in addition to changing the economic situation, can also undermine occupational stability and contribute to a breakdown in the family. Similarly, impacts can be positive *and* negative. In other words, they have pros and cons.

The pros of rural tourism development include the acquisition of additional income through the leasing of land, the letting or sale of buildings, the sale of products, employment in local tourist activities and services, and the growth of pride and sense of place (Mathieson and Wall, 1982; Ryan, 1991). The negative impacts include the creation of conflicts in the countryside (e.g., between tourists and local people, and among local people themselves), increased erosion, increased competition for land, and the 'tracking' of young people into low-level jobs in the service sector.

Gannon (1994) gives a good overview of the pros and cons of rural tourism development (Table 10.1). The intensity and effect of rural tourism impacts may vary from situation to situation. To understand how this is possible and to get a complete picture of the consequences of development, we must consider not only which impacts are produced by which types of tourists, but also on which groups in the host community they have the most impact. Dogan (1989) and Ryan (1991) pointed out that host communities are not necessarily homogeneous, and the interests, attitudes, and perceptions of local residents towards rural tourism development and its impacts can vary. Studies show that only a small proportion of the host population usually benefits from tourism, and thus attitudes towards tourism in the host community tend to be more favourable among residents who depend on it for their livelihood (Milman and Pizam, 1988; Pizam, 1978).

Table 10.1: The pros and cons of rural tourism development

Economic costs Rural tourism: • Incurs developmental and marketing costs. • Involves risk of the operator and community. • Places demands on public services. • May only give part-time employment. • Is an image industry and is therefore very sensitive to the macro environmental conditions and forces outside direct community control. • May increase cost of living for community residents because of inflation of property values and goods and services costs.	**Economic benefits** Rural tourism: • Brings more money, particularly foreign currency, to the economy. • Creates jobs and increases family and community income. • Helps to diversify and stabilize the rural economy. • Provides broader business base for the community and creates an opportunity for attracting other businesses and small industries. • Provides the opportunity for innovation and creativity. • Provides the support for existing businesses and services. • Helps to develop local craft and trade.
Social costs Rural tourism may cause: • The introduction of conflicting ideas and styles into the community. • The increase of crime. • Overcrowding and congestion. • Infringement on privacy in households and in the sharing of important community resources with outsiders. • Institutional, community and family jealousies, as all may not share the benefit equally.	**Social benefits** Rural tourism: • Fosters a pride of place and sense of place. • Helps to build up a community infrastructure. • Provides the opportunities for cultural exchange. • Promotes a team community spirit particularly through the development of cultural and entertainment activities. • Creates conditions for safeguarding and enhancing local cultural identities.
Environmental costs Rural tourism: • May downgrade the quality of natural and historic areas through increased tourism development and tourist behaviour. • May increase noise and litter pollution.	**Environmental benefits** Rural tourism: • Can be a key factor in revitalizing the natural, cultural and historical resources of a rural region. • Promotes and encourages village renewal and a cleaner countryside. • Can foster the preservation and conservation of worthwhile resources.

Source: Gannon (1994)

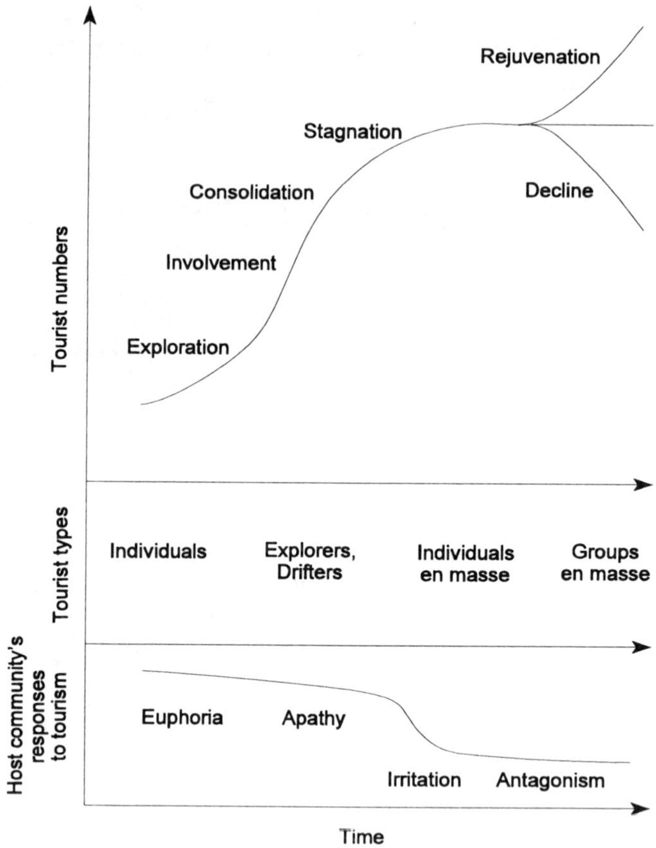

Figure 10.2: The host community's responses to rural tourism throughout the life cycle of a tourist destination area (after Butler, 1980; Butler and Waldbrook, 1991; Doxey, 1975)

Figure 10.2 illustrates the correlation between the responses of local residents to tourists and the impacts of tourism. Butler (1980) identified six stages in his cycle: exploration, involvement, development, consolidation, and stagnation, followed by either decline or rejuvenation. In the first stage of the cycle (exploration) the local people are enthusiastic and thrilled by the development of tourism. In the second stage (involvement) they are blasé. Local control and euphoria decline rapidly in the third stage (development). At this point, it is likely that local residents are irritated by the growing numbers of tourists that impose on their lives. In the fourth stage (consolidation) external forces move in (e.g., investors, bigger companies). The local people become antagonistic in the fifth stage (stagnation). In the sixth stage there are two possible scenarios. The first of these (rejuvenation) signals the beginning of a new boom in rural tourism development. There are, however, very few examples of rejuvenation in the literature. The second (decline) is the more likely outcome. In this scenario, the local people have to learn to live with the fact that their environments and lives will never be the same, even though they may regain control over their resources. In most cases, the local people and their natural settings are often no better off after the introduction of tourism than they were before.

Is it possible to avoid decline, for the sake of the local communities and the tourists? Unfortunately, as Butler (1993) pointed out, there is little recognition among the stakeholders involved in rural tourism development that the tourist industry behaves like most other industries, namely that it can be self-perpetuating and not easily reversible. There is also little agreement among these stakeholders on desirable levels of rural tourism development, and even less on the direction it should take. Too often rural tourism is planned from 'outside,' and the needs and welfare of the host communities are neglected.

Rural tourism development has the power to change much in rural society and the environment. While we acknowledge that some of these changes may be simply the result of any kind of economic and social development, we cannot deny that the growth of rural tourism raises many questions about the social and environmental desirability of encouraging further expansion. There is a need for the kind of rural tourism development that maintains and revitalizes the countryside and rural societies while it preserves the environment. In such a context, sustainability becomes an important issue.

Sustainability: the way forward for rural tourism development?

What do we mean by sustainability? Although there are as many definitions of the term as there are people writing about it, the essence of the debate is always how to safeguard nature's productive capacity to support human existence in a way which best fits the lifestyle aspirations of current and future generations (WCED, 1987; Campbell et al., 1993).

For tourism development in general, Jenkins (1991) defined sustainability as 'the careful use of current tourism assets to ensure their future availability.' This definition seems to be narrow, as it fails to acknowledge that sustainability is about more than preserving tourism assets. Inskeep (1991) gave a much more detailed definition of sustainable tourism development: "Sustainable tourism development aims to protect and enhance the environment; to meet basic human needs; to promote current and interregional equity; and to improve the life of all people by helping to integrate conservation, cultural compatibility, and local input." Inskeep underlines the importance not only of the environmental and developmental components of sustainability, but also of the active involvement of the host community in tourism development.

In the context of rural tourism development, Jones (1993) pointed out the importance of the local community's participation. He stated that sustainable (rural) tourism development should respect the needs and interests of the host population and support local control over decision making. It should encourage interaction between hosts and guests, and promote cultural development (including the choice of lifestyle and preservation of local traditions).

I believe strongly that sustainable rural tourism development can lead to a balanced interaction between the major 'ingredients' of rural communities, namely the economy, the environment, and society. I do not believe, however, that we can ignore the pitfalls of rural tourism development. Therefore, any discussion of sustainable tourism development should include such questions as: What do we wish to sustain and for whose benefit? Which parameters do we measure, over how large an area, and for how long?

The answers to these questions depend on the goals, values, and priorities of the different stakeholders involved in rural tourism development. This means that assessments of sustainability are relative and socially constructed. Thus, although it is possible to define and measure the social, economic, and ecological parameters of tourism development (such as carrying capacity, saturation level), an evaluation of them will always be relative and a matter of negotiation. Certain alternative trends (such as 'ecotourism,' 'green tourism,' 'soft tourism') are gaining in preference to mass tourism. These, too, can be 'more sustainable' or 'less sustainable,' 'sustainable' or 'non-sustainable.' As Wight (1993) pointed out, mass tourism may not always be undesirable or inappropriate, and small-scale tourism may not always be sustainable. An assessment of sustainability, therefore, must take into account the specific context in which rural tourism development takes place, including the economic, environmental, social, and political components.

A Slovenian case

Slovenia is one of the youngest countries in Europe.[4] In the five years since gaining its independence, Slovenia has been actively creating a niche for itself in the New Europe. The government has been formulating new policies on all aspects of national life, including rural development and tourism. The policies for both of these sectors emphasize the importance of the countryside and rural communities as potential resources for, and beneficiaries of, tourism development, although perhaps not for the same reasons. The tourism planners view rural areas as a part of the total tourist product and are interested mainly in tourism's contribution to the national economy. The rural development planners are concerned about tourism's positive benefits for rural communities. How ready are Slovenia's countryside and its rural societies to face the challenges and impacts of rural tourism development?

Slovenia covers an area of 20,256 km^2. The population numbers approximately 2 million, with a density of 93.4 people per km^2. The country is mountainous, being dominated by the Julian, Karavanke, and Kamnik ranges of the Eastern Alps. It has a stretch of coastline on the Adriatic Sea (46 km). It is bordered by Italy on the West, Austria on the North, Hungary on the East, and Croatia on the South. The capital, Ljubljana, has a population of 305,000. Ljubljana is also the administrative and cultural centre of the country. The variation in climate and geography is great, offering possibilities for the development of different types of tourism in the Alpine, Adriatic, Karst and Panonian regions.[5]

Key features of the economy

Slovenia is relatively highly industrialized and export oriented. With a per capita income of US $ 10,000 (Delo, 1996), it ranks among the moderately developed countries – behind Spain and Ireland, for example, but better off than Greece, Portugal, and the Eastern and Central European countries. Most of the country's gross domestic product is generated by industry (37.8 per cent), agriculture (4.9 per cent), and trade and business services (57.8 per cent). The low share of tourism (2.9 per cent) is owing to the decrease in the number of foreign visitors since 1990. The main factors responsible for the decline in tourism are the atmosphere of extreme uncertainty caused by the war in the Balkans, the limited outside awareness that Slovenia is an independent state, the confusion caused by a name with several variations (Slovakia, Slavonia, Slovenia), and the lack of distinct image. But despite these problems and the current underdevelopment, tourism is becoming ever more important. The expectation is that it will play a considerable role in Slovenia's future economic structure. The potential for tourism in Slovenia is based on the country's outstanding historical and cultural heritage, its natural resources (most of them still in pristine condition), a temperate climate, and a favourable geographic location. Different types of tourism have developed already, many of them in the countryside. These include health tourism, winter sports, cultural tourism, social tourism, youth tourism, and rural tourism.[6]

The current situation in the countryside

Rural areas and agriculture are important to the Slovenian economy, even though agriculture accounts for only 4.9 per cent of the total GDP. The main types of land use are agriculture and forestry. Of the country's total area, 42.7 per cent is used for agriculture, and 50.2 per cent is covered by forests (Table 10.2). Approximately 80 per cent of total agricultural land is privately owned (National Agency for Statistics, 1994). According to the 1994 census, almost half of the population lives in rural areas. Of the total working population, 11 per cent are engaged directly in agriculture and forestry,[7] and more than 30 per cent work in related industries (e.g., food-processing industry, rural tourism, and so on).

Table 10.2: Surface and land structure in Slovenia, 1960–1990 (1,000 ha)

	1960	1970	1980	1990
Total Agricultural land*	937.6	944.9	875.2	866.4
Fields, gardens	294.7	275.0	253.9	247.0
Orchards	29.2	35.8	36.1	36.3
Vineyards	23.1	20.6	21.3	21.5
Meadows	288.0	323.4	333.4	347.7
Pastures	299.2	286.8	226.8	211.5
Forests	958.4	949.6	1014.0	1024.5

*In Slovenia agricultural land includes cultivated land (fields, orchards, meadows etc.) as well as pastures.

What these data do not illustrate is the real situation in Slovenia's countryside. Almost half of the population (nearly 1 million people) live in the countryside, but only 7.6 per cent are farmers. The number of farmers decreased by 11.5 per cent (some 69,000) in the last decade. The unfavourable age structure and intense emigration are the main reasons for this downward trend. Moreover, there are increasingly fewer farms where all the members of household engage exclusively in agriculture. More and more of the rural youth are taking jobs outside agriculture. Agricultural production is being taken on by pensioners (25 per cent) and part-time farmers (20 per cent). In 1991, only 12 per cent of the farmers were working full time on their own holdings, earning their total income on the farm. More than 70 per cent of farmers were part-time farming.[8]

Socio-economic farm structure

In 1994 there were 156,549 farms in Slovenia.[9] Almost one half of them were classified as *part-time farms*, that is, farms with at least one member of household employed elsewhere. There were 21.3 per cent farms that were defined as *full-time farms*, that is, where the entire household works on the farm only, or is supported by those who work on the farm only. It is necessary to stress here that socio-economic structures differ amongst the farming regions of Slovenia. The percentage of full-time farms is much higher in the mountain regions. The share of full-time farms in the Karst, however, is under the average, as this region has better access to employment centres.

The average size of a Slovenian farm is 7.8 ha. This area includes both agricultural and forest land. Almost one-third of all farms (28.4 per cent) has less than 1 hectare of land. The largest group (36 per cent) comprises the farms that have between 1 and 5 hectares, with only 17.6 per cent having a total area bigger than 10 ha (Kovačič, 1995). In recent years, the average farm size has decreased even though there has been similar reduction in the number of farms. Data show that less than 25 per cent of all agricultural land is in the hands of the supplementary and aged farmers (those older than 64 years). The part-time farmers hold almost half of the total land.[10]

Livestock production is the predominant branch of Slovenian agriculture. Its prominence is the legacy of the government's agrarian development strategy of the early 1970s. The overproduction of milk and meat during those years caused market disparities, with low production of all other basic agricultural products. To this day, self-sufficiency is important on farms, especially on the supplementary ones, but also on a significant number of part-time farms.

Rural tourism development

Rural tourism in Slovenia goes back to the nineteenth century. By the first half of the twentieth century it was well established. A real boom in rural tourism development occurred in the late 1970s, fuelled in part by the government's growing concern about securing additional income for mountain farmers, and helped along by the Agricultural Advisory Service, which trained special advisors for farm families, and for rural tourism development.

Over the years, different forms of rural tourism have emerged, the most important being agri-tourism, that is, tourism on family farms, where guests stay either with the farm family or in a guest house, and can choose accommodation ranging from bed and breakfast to full room and board. Agri-tourism is often referred to as 'stationary rural tourism.' A second form, known as 'trip tourism,' is based on 'open-door farms,' where visitors can have a meal and explore the farmyard for just a few hours. A third, very recent, form allows visitors to pitch their tents and camp at the farm. Of these three, agri-tourism has been the most important for farmers and visitors to rural areas.

Over the last twenty years, agri-tourism has expanded all over Slovenia, primarily to provide a secondary source of income for family farm households (Jemec, 1990). Robič (1989) points out that Slovenian farmers, like other European farmers, have been forced to adapt to worsening circumstances (low income, small farm sizes, limited production conditions, and continuing depopulation trends). In 1990 there were 160 agri-tourism farms, almost all of them with rooms of the first and second categories, that is, having their own shower and toilet. In the last five years, another 300 farms have involved themselves in rural tourism development by joining either the projects of the Centre for Rural Development and Revitalization of Villages (CRPOV), or those of the cooperatively owned tourist agency VAS. Advice and infrastructural support for rural tourism activities come from such organizations as the Agricultural Extension Service, established within the Ministry for Agriculture and Forestry, the Centre for Rural Development and Revitalization of Villages, and VAS, which was established in 1990 by farmers, and has in the last few years successfully marketed and promoted many of the agri-tourism farms through their own agency. But despite these efforts, agri-tourism is not gaining in strength evenly everywhere in Slovenia.

Significant sums in the form of special loans and grants have been given to promote the development of rural tourism. In many cases, however, these investments have not worked out. Farms with tourism have been forced to turn into inns to attract enough guests to return the money they received. This is especially true in the areas with a poorly developed infrastructure. Research in the Slovenian countryside (Markovič, 1991; Verbole, 1993, 1994) shows that farmers – and farm women – do not always have positive attitudes toward rural tourism. Many found tourists too demanding, wanting luxuriously furnished rooms the farmers are not able to provide, and food that is rarely served at the farmers' own table, a phenomenon also known as the 'ecological bubble.' Farm women complained that catering for tourists put too much extra pressure on them. Many of them shared the opinion: 'Tourism? No thank you!' Many farmers reported that tourists damaged the farmland and mistreated the farm animals. Especially when the farms were close to urban areas, there were increasing problems of trespassing, dumping of rubbish and harassing of livestock, which is illustrated by the following views expressed by farmers (Verbole, 1993):

Livestock and tourism do not go well together (Farmer Ober, 1992).

I do not like those 'city' tourists. Last summer there was a group camping (wild-camping) on my meadow and all they left behind them was garbage (Farmer Jože, 1992).

. . . they came with their motorbikes and were running after my calves. One broke a leg and the vet had to put the animal down (Woman farmer Marija, 1992).

For the last two or three years, they (the local Tourists Office) have been organizing mountain bike tours through the 'untouched' nature of the park. And then, what happens? You see them riding down the meadows damaging grass turf. Somebody should tell them that this is damaging to the soil (Farmer Janez, 1992).

But rural tourism also brings along with it many positive experiences. Some of these are the following: new house can be built thanks to the extra income; children make friends with the guests and spend the holidays with them (even abroad); the farmers learn to speak (or get the opportunity to practise) a foreign language; farm women attend courses organized by the Agricultural Advisory Service on cooking and nutrition, how to decorate food, how to do efficient bookkeeping; farmers visit other farms where tourism is prospering; they go on trips together (Verbole, 1996b).

The official Slovenian Strategy of Tourism Development calls for rural tourism development to be expanded. As Slovenian tourism development in general is at a delicate stage now, it is vital to decide which of the following alternatives will be the future of the Slovenian countryside: one big resort for mass tourism; a sort of compact, alpine, open-air museum for Japanese and American tourists; a viable national resource, responsibly used and managed, and accessible through sustainable tourism?

Slovenia's agricultural sector and rural areas have undergone dynamic changes since the country's independence. The most significant of these are the changes that followed in the wake of democratization, decentralization, and privatization (Mešl, 1996; quoted in Verbole, 1996a).

Democratization has directly influenced the national political, economic, and social systems. The change in the distribution of power is evident in the new governmental structure, the shift in policies, and the adjustments to existing structures. Democratization has enabled stronger involvement of the Slovenian Farmers' Party (Slovenska ljudska stranka) in agricultural policy developments. Consequently, several of the changes in these developments are because of the pressure brought to bear by the farmers' lobby. In 1992, the new agricultural development strategy gained parliamentary approval. The primary goals are to assure the stable production of good-quality food at reasonable prices and the food security of the country, to keep the countryside populated, to protect and preserve the cultural landscape, agricultural lands, farms, and water resources, to increase the market competitiveness of agricultural produce, and to generate reasonable income for productive farmers. One year earlier, in 1991, a special law on revitalizing the development of depopulated areas was passed as part of the new development-support measures package that had been put together to provide people in the countryside with information and knowledge (local development plans), infrastructure (investment grants), and start-up capital (loans to stimulate production). Additional opportunities and challenges for investors in the rural sector were opened by the passage of the new law on enterprises, which names private initiative as a vehicle for economic development in all sectors, including agriculture. Several additional support measures for entrepreneurship in rural areas have followed since then.

Decentralization has given the different regions more say in the decision-making process, stimulated the restructuring of the local government,[11] and provided the opportunities to formulate regional development strategies. In theory, decentralization means more equal distribution of power and direct support for local development (budgeting). It also means more opportunities for new programmes and support measures for local development (such as the national programme for rural development and village renewal of the CRPOV).

Privatization has been the basis of Slovenia's economic restructuring. The national law on privatization has created new conditions, legislation, and structures for the rural sector. Under the law, many of the forests that were formerly state-owned have been denationalized. Some of the forest land has been given back to the original owners, only a small number of whom were farmers.[12] Privatization and related new legislations have destroyed the monopoly of agricultural cooperatives on the marketing of agricultural products. The establishment of the National Land and Forest Fund has created the basis for a market in land.

The hope has been that the structural reforms in the rural sector will create favourable conditions for solving the problems of size structure (opening a land market), socio-economic structure (new opportunities for economic activities in rural areas, such as rural tourism), and market conditions (new opportunities of generating income). Before this hope becomes a reality, however, the planners will have to deal with the challenges and potential problems arising from the new conditions in the rural sector, including rural tourism development.

The main challenges are the following: how to stimulate the development of new activities such as tourism on farms and in rural areas within the given (ageing) population structure and the constraints of low investment ability and a poor educational structure; how to stimulate active participation of local communities in the new tourism development projects under the national programme for rural development and village renewal of the CRPOV; how to ensure that local communities and their environment suffer the minimum costs and receive the maximum benefits of tourism development.

Conclusions

Raising the quality of life is a precondition for maintaining the viability of rural areas. As agriculture can no longer be the sole economic basis for long-term development in the countryside, it is vital to put more consideration into planning new kinds of development and improved techniques for managing them. There is a need for a new, more global, and comprehensive approach to formalize and install new orientations into official policies for rural development.

One of the options for the development of rural areas as a whole is rural tourism. As such, rural tourism and its development can not be separated from the context of rural development. This means rural tourism development must be approached critically, and a key criterion for its implementation must be that it should contribute to the sustainable development of rural areas.

Rural tourism development can, as pointed out throughout this Chapter, have not only positive, but also negative impact on rural areas. To minimize the negative impact, rural tourism development should aim at protecting and enhancing the environment, promoting interregional equity, improving the quality of life by integrating conservation, cultural compatibility and local input. In other words, it should aim at being sustainable.

It will only be possible to achieve sustainable tourism development in rural areas of Slovenia by maintaining the unique character of the host communities and their environment. This means that traditional rural tourism development, based on bringing in programmes and capital from outside, cannot be the right approach. Sustainable rural tourism development in Slovenia should, therefore, be based on defining and activating local development to realize the indigenous potentials of rural areas in all their geographic, cultural, and natural diversity. Once in place, this approach will require coalitions and linkages between the different stakeholders – the state, institutions, and local communities – involved in rural development.

Acknowledgement

The author wishes to thank Dr. P.P.A. Giparaki for his kind comments on a draft of this Chapter, and Mrs. M.M. Naeff-Snyder for editing the final version.

Notes

1. The former social benefits in Slovenia, for example, included free education for everybody from primary school up to university, free health care, guaranteed pensions for everyone including farmers, paid maternity leave for one year for all mothers (optional for fathers) including, since 1982, farm mothers, and cheap holidays for every worker ('social tourism').
2. Rural tourism in the last three decades has been characterized by a growing number of tourists. Expectations for the years to come are that the demand for tourism in the countryside will rise steadily, in accordance with the general upward trend in tourism (Hummelbrunner and Miglbauer, 1994).
3. Kovacs, 1993
4. Slovenia was declared an independent state on 25 June 1991 and recognized internationally on 15 January 1992.
5. The climate varies from Mediterranean on the coast, to Alpine in the North, to Continental in the East.
6. The future tourism offer of Slovenia concentrates mainly on the natural and geographical features of the country. Proposals for further tourism development mention four touristic 'clusters': the Alpine cluster, which includes winter and summer activities; the health resorts cluster; the Coast and Karst cluster; the cities cluster.
7. This number includes the active farm population and hired workers.
8. Slovenian statistics differentiate between part-time mixed farms, that is, farms where at least one member of the farm family is fully employed on the farm, and part-time supplementary farms, i.e., farms where all of the family members are employed off farm.
9. This number has decreased by 4.5 per cent over the last decade (Kovačič, 1995).
10. This implies that potential land reserves are relatively modest.
11. The concept of local self-government is related to the separation of the functions of the state from those of local self-governing bodies. In this situation, the municipality becomes a local self-governing community (Vlaj, 1995).
12. Before 1945, many of Slovenia's national forests belonged to the Catholic Church and to Austrian and German aristocrats, great landowners, and traders. Bowing to public pressure, parliament passed a special act in 1995 to freeze the denationalization of Church-owned forests.

References

Barbič, A. (1991), *Kmetov vsakdan: položaj in prihodnost družinskih kmetij na Slovenskem,* Cankarjeva založba; Ljubljana.

Barbič, A. (1992), 'Rural Development in the Time of Deconstructing the One-Party Political Systems and Centrally Planned Economies,' Paper presented at the 5th ESRS Summer School, Trondheim, Norway, June 13.

Barbič, A. (1993), 'Sustainable Diversification – the Future for Rural Areas: The Case of Post-Socialist Countries,' *Zbornik Biotehniške fakultete Univerze v Ljubljani*, No. 61, pp. 209–222.

Bramwell, B. (1994), 'Rural Tourism and Sustainable Rural Tourism,' *Journal of Sustainable Tourism*, Vol. 2, No. 1/2, pp. 1–7.

WCED – World Commission of Environment and Development (1987), *Our Common Future*, Oxford University Press: Oxford/New York.

Butler, R.W. (1980), 'The Concept of a Tourist Area Cycle of Evolution: Implications for Management of Resources,' *Canadian Geographer*, Vol. 24, pp. 5–12.

Butler, R.W. (1993), 'Pre- and Post Impact Assessment of Tourism Development,' in D.G. Pearce and Butler, R.W., *Tourism Research: Critiques and Challenges*, Routledge: London and New York.

Butler, R.W. and Waldbrook, L.A. (1991), 'A New Planning Tool: The Tourism Opportunity Spectrum,' *The Journal of Tourism Studies*, Vol. 2, No. 1, pp. 3–14.

Campbell, A. et al. (1992), *The Environmental Impact of the MacSharry Reforms on EC Agricultural Policy*, Department of Ecological Agriculture, Wageningen Agricultural University, The Netherlands: Wageningen.

Cloke, P. and Goodwin, M. (1992), 'Conceptualizing Countryside Change: From Post-Fordism to Rural Structured Coherence,' *Transactions of the Institute of British Geographers*, Vol. 17, pp. 321–336.

Dann, G. and Cohen, E. (1991), 'Sociology and Tourism,' *Annals of Tourism Research*, Vol. 18, pp. 155–169.

Dernoi, L.A. (1983), *Present Conditions of Farm Tourism in Europe*, Agrarian Structures and Farm Rationalisation Report No. 2, FAO/ECE/GRI/WP3/12, United Nations: New York.

Dogan, H.Z. (1989), 'Forms of Adjustment: Sociocultural Impacts of Tourism,' *Annals of Tourism Research*, Vol. 16, pp. 216–236.

Doxey, G.V. (1975), 'A Causation Theory of Visitor-Residents Irritants; Methodology and Research Inference,' *Proceedings of the Travel Research Association*, 6th Annual Conference, San Diego, California.

Gannon, A. (1994), 'Rural Tourism as a Factor in Rural Community Economic Development for Economies in Transition,' *Journal of Sustainable Tourism*, Vol. 2, No. 1/2, pp. 51–61.

Giagou, D. (1996), 'Agritourism in Greece,' Paper presented at the 7th ESRS Summer School, Veszprem, Hungary, July 1–5.

Hummelbrunner, R. and Miglbauer, E. (1994), 'Tourism Promotion and Potential in Peripheral Areas: The Austrian Case,' *Journal of Sustainable Tourism*, Vol. 2, No. 1/2, pp. 41–51.

Inskeep, E. (1991), *Tourism Planning: An Integrated and Sustainable Development Approach*, Van Nostrand Reinhold: New York.

Jansen-Verbeke, M. (1990), *The Potentials of Rural Tourism and Agritourism in the EEC*, University of Nijmegen: Nijmegen Netherlands.

Jansen-Verbeke, M. (1991), *Towards Balanced Tourism Development in Europe*, University of Nijmegen: Nijmegen

Jemec, M. (1990), 'Introduction to Agritourism in Slovenia,' Handouts of the IAAS seminar Slovenia's Mountain Farming, Bled.

Jenkins, C.L. (1991), 'Development Strategies,' in Lickorish, L.J. et al. (eds), *Developing Tourism Destinations*, Longman Group UK: Hong Kong.

Jones, A. (1993), 'Sustainability and Community Participation in Rural Tourism,' *Leisure Studies*, Vol. 12, pp. 147–148.

Keane, M. (1992), 'Rural Tourism and Rural Development,' in Briassoulis, H. and Straaten J. van der (eds), *Tourism and the Environment: Regional, Economic and Policy Issues,* Kluwer Academic Publisher: Dordrecht

Keane, M.J. and Quinn, J. (1990), *Rural Development and Rural Tourism,* Research Report No. 5, Social Sciences Research Centre, University College Galway: Galway.

Kloeze, J.W. te (1994), *Agritourism as Reproduction of Human and Spatial Agrarian Resources,* Centre for Recreation and Leisure Studies, Wageningen Agricultural University: Wageningen.

Kovacs, D. (1993), 'Emerging Rural Tourism in the Changing Rural Society of Hungary,' Paper presented at the XVth European Congress of Rural Sociology, Wageningen, The Netherlands, 5– 11 August.

Lane, B. (1994), 'Sustainable Rural Tourism Strategies: A Tool for Development and Conservation,' *Journal of Sustainable Tourism,* Vol. 2, No. 1/2, pp. 102–111.

Markovič, M. (1991), 'Družine z vitalnih in odmirajočih kmetij v luči nekaterih odnosov,' In Barbič, A. (ed.), *Prihodnost slovenskega podeželja: prostor, prebivalci, gospodarske dejavnosti*, Tiskarna Novo mesto: Novo mesto.

Mathieson, A. and Wall, G. (1982), *Tourism: Economic, Physical and Social Impacts*, Longman Scientific and Technical Publishers: London.

Milman, A. and Pizam, A. (1988), 'The Social Impact of Tourism on Central Florida,' *Annals of Tourist Research*, Vol. 15, pp. 191–204.

Murphy, P. (1985), *Tourism: A Community Approach*, Methuen: London.

Pearce, P.L. (1990), 'Farm Tourism in New Zealand: A Social Situation Analysis,' *Annals of Tourism Research*, Vol. 17, pp. 337–352.

Pizam, A. (1978), 'Tourism Impacts: The Social Costs to the Destination Community as Perceived by Residents,' *Journal of Travel Research,* Vol. 16, No. 4, pp. 8–12.

Ryan, C. (1991), *Recreational Tourism: A Social Science Perspective,* Routledge: London and New York.

Urry, J. (1990), *The Tourist Gaze: Leisure and Travel in Contemporary Societies*, Sage Publications: London.
Verbole, A. (1993), *Slovenian Mountain Farmers and Their Role in Environmental Conservation: A Case Study of the Interface in the Triglav National Park*, MSc Thesis, Wageningen Agricultural University: Wageningen.
Verbole, A. (1994), 'Tourism Development in the European Countryside: Costs and Benefits,' Paper presented at the International Workshop Residents - Management - Sustainable Development, Slovenia, Pišece, 25–30 June.
Verbole, A. (1995), 'Negotiating the Terms of Tourism Development in the Countryside: Political Considerations in Sustainable Rural Tourism Development,' Paper presented at the international workshop 'Rural communities: local responses to global questions,' Eisenkappel, Austria, October 12–14.
Verbole, A. (1996a), 'Approaching EU?: Slovenia's New Rural Development Strategies,' Paper presented at the 7th ESRS Summmer School, Veszprem, Hungary, July, 1-5.
Verbole, A. (1996b), 'Making Women Visible: The National Action Plan for Rural Women in Slovenia,' Paper presented at the 9th World Congress of Rural Sociologists, Bucharest, Romania, July 22–27.
Wight, P.A. (1993), 'Sustainable Ecotourism: Balancing Economic, Environmental and Social Goals within an Ethical Framework,' *The Journal of Tourism Studies*, Vol. 4, No. 2, pp. 54–67.